Getting Started with Beef & Dairy Cattle

GETTING STARTED
with
Beef & Dairy Cattle

Heather Smith Thomas

Illustrations by James Robins

Storey Publishing

The mission of Storey Publishing is to serve our customers by publishing practical information that encourages personal independence in harmony with the environment.

Edited by Deb Burns, Sarah Guare, and Kathy Casteel
Cover and text design by Kent Lew
Text production by Jessica Armstrong and Jennifer Jepson Smith
Cover photograph by © Arthur C. Smith III / Grant Heilman Photography
Spine photograph by © Denny Eilers / Grant Heilman Photography
Illustrations © James Robins
Photo page vi courtesy of Heather Smith Thomas
Indexed by Christine R. Lindemer, Boston Road Communications

Printed in the United States by Versa Press
10 9 8 7 6 5 4 3 2 1

Library of Congress Cataloging-in-Publication Data

Thomas, Heather Smith, 1944–
    Getting started with beef & dairy cattle / Heather Smith Thomas.
        p. cm.
    Includes bibliographical references and index.
    ISBN-13: 978-1-58017-596-8; ISBN-10: 1-58017-596-1 (pbk.: alk. paper)
    ISBN-13: 978-1-58017-604-0; ISBN-10: 1-58017-604-6 (hardcover: alk. paper)
1. Beef cattle. 2. Dairy cattle. I. Title: Getting started with beef and dairy cattle.
II. Title.

SF207.T46 2005
636.2'1—dc22

                                                                       2005009458

Contents

This book is dedicated to all the unique and special cattle

we have raised over the years, and especially

to those wonderful creatures that became our "pets,"

giving our family much enjoyment.

Introduction

RAISING CATTLE — A MILK COW, or a calf for freezer beef — can be a profitable and satisfying experience. Even with limited time, you can still find ways to keep a milk cow, sharing her milk with her calf (or her own calf and an extra one), milking her only when you need the milk and letting the calves nurse her the rest of the time. Or, if you enjoy dairy cattle, you may wish to have several cows to operate a small dairy.

The wonderful thing about cattle is that they are very efficient food producers, converting grass into meat or milk. They can eat roughages that humans cannot, grazing on land that won't grow crops. They can mow the hillside behind the house that is too steep for gardening or field crops, or graze the "back forty" that is too brushy, rocky, or swampy to grow anything other than patches of grass. Cattle can provide us with meat or milk while keeping the weeds trimmed, providing fire control and a neater landscape. If you have enough acreage to raise a few extra animals to sell meat or milk, cattle are a way to harvest the grass and other forages and convert these plants into a salable product.

In many situations there is not much expense involved, except for the initial fencing to keep them where you want them. If you live in a moderate climate where grass grows for most of the year and you don't need to provide much supplemental feed, you can raise cattle very reasonably. Unlike poultry or hogs, cattle can do well on forages alone. You can raise grass-fat beef without any grain. Even in regions where snow covers pastures in winter people can still raise beef animals without hay. Weaned calves can be purchased in spring to harvest grass during the growing season, then butchered or sold in the fall before it's time to feed hay. Yearling cattle are a very efficient way to make use of seasonal pasture, to make a little extra money from an acreage, or to create a harvest of meat.

Raising your own beef gives you the satisfaction of knowing its history. Not only do you have the pleasure of knowing the animal personally, but you also have control over how he is raised and what he eats. You can grow beef without the use of antibiotics, growth hormones, insecticides, or other products that you may wish to avoid. When you raise your own beef, you know exactly how he was fed and the degree of care that is taken. There may be instances in which you must give an animal antibiotics in order to treat a certain illness, but if you take good care of the animal and he has a clean and healthy place to live, the risk of illness is reduced.

Raising cattle is rewarding, entertaining, and soul-satisfying. Our lives are touched by these animals, and we can appreciate our personal relationship with them. They are fascinating animals; working with cattle is never boring. Life with cattle can be physically challenging at times — as when trying to deliver a calf in a difficult birth or trying to catch an elusive calf! For those of us who enjoy cattle, the chores of caring for them are not work. This "work" is our pleasure; cattle add more quality of life to our daily experience.

Just as a gardener takes satisfaction in producing fruits and vegetables for the family larder, the cattle caretaker enjoys providing the meat or milk. Humans and cattle have had a very intimate interrelationship for thousands of years. Our prehistoric ancestors first hunted them, then caught and tamed them about 10,000 years ago. Cattle became a handy source of meat and hides. Later, we learned to hitch them to a cart or plow; oxen were used for transportation long before horses were.

There is a distinct naturalness about raising cattle. It's good for the soul. In tending a garden or caring for animals, we find harmony with the earth and a purpose in our lives. The person who grows his or her own food takes on a responsibility that enlarges his or her awareness of life and one's own part in it. Animal agriculture — working with the land and the animals the land supports — helps us find our own "fit" in the whole of creation.

This book will be a handy reference for anyone who wants to raise a calf or a whole herd of cattle. It is written in simple terms so even the beginner or young stockman can understand it. At the same time, it covers a wealth of information that will be useful to anyone who has beef or dairy cattle and can serve as a cattle manual for future reference. It will also be of help to parents whose children have 4-H or FFA projects. Most cattle caretakers need advice from time to time from a veterinarian, cattle breeder, dairyman, or county Extension agent. Don't hesitate to contact an experienced person to request help.

CHAPTER ONE

What Do I Need?

RAISING CATTLE CAN BE FUN and a rewarding challenge. Their versatility gives you several options in housing and feeding methods. You can raise a steer in a corral or on a small acreage, or you can keep a herd of cattle on a large pasture, on crop stubble after harvest, or grazing steep, rocky hillsides.

Cattle can be fed hay, or hay and grain, or fed entirely on forages they harvest themselves. Your climate, land, economics, and individual circumstances will dictate your methods. Before you bring animals onto your place, determine the facilities, fencing, and feeding method you can provide. If you have good pasture land, all you need is proper fencing to keep the animals in it.

What Kind of Animals Should I Raise?

You have several choices when it comes to deciding what type of cattle you want, or how many. Your choices will be influenced by how much space you have and what your goals are. You can raise a steer or two for beef, a heifer (beef or dairy) to become a cow, or a small herd of cows for raising calves.

If you are raising beef calves to sell, raise steers. They bring more money per pound than heifers when sold, and also weigh more than heifers of the same age. If you want to keep the calves, choose heifers. You can raise beef heifers to become cows (to raise more calves to sell), or a dairy heifer to become a family milk cow, or several heifers for a small dairy. You may also want to raise dairy heifers to sell; dairy heifers are worth more than beef cattle when they grow up.

Recognizing Males and Females

If you have no experience with cattle, you'll need to learn the difference between a steer and a heifer. Bulls and steers are males. When a male calf is born he is a bull. His reproductive organs are his testicles and penis. His testicles are inside his scrotum — the sac-like pouch that hangs between his hind legs. His penis is inside his sheath (a tubular fold of skin), located on the underside of his belly.

Most bulls end up as steers. A bull becomes a steer when his testicles are removed in a process called castration. The steer may still have a scrotum (though it will be small and empty), or his scrotum may be completely gone, depending on the method used to castrate him (see page 91). Only the best males are kept as bulls for breeding with cows.

Most ranchers castrate all bull calves to sell as steers, and buy bulls from a seed stock breeder who raises the breed of cattle they want (either purebreds or composites; the latter are a mix of two or three breeds to combine best qualities of each). People who raise beef cattle to butcher or to sell to a cattle buyer castrate the male calves because

bull   steer

heifer

cow

they are more docile and may gain weight better because they spend less energy fighting or trying to chase after cows.

A heifer is a young female. Most of her reproductive organs are inside, so you can't see them. A heifer calf has a tiny udder between her hind legs, with little teats. A bull or steer calf also has small teats, just as a boy has nipples, but they don't grow. The heifer urinates from an opening by her vulva, which is located under her tail below the rectal opening. The vulva is the opening to reproductive organs located inside her body.

What Breed Should I Choose?

There are dozens of breeds of beef cattle and several dairy breeds to choose from. Your choice might be influenced by color of the breed, its unique characteristics, its interesting history, popularity in your area, or the fact that friends or acquaintances raise that breed.

Once you decide on the breed, have a knowledgeable person help you pick out the animals. You want to make sure you choose good ones that will stay healthy and perform well. An experienced cattle person can help you make better decisions regarding the health and desirable characteristics of the animals you might buy.

There are outstanding animals in every breed, and many good crossbred animals. If you want to raise beef cattle, crossbreds will generally outperform purebreds in almost all desired traits, due to a factor called hybrid vigor. They often gain weight faster, are more feed efficient, and are more fertile for raising calves. If you want to raise dairy heifers to sell, however, choose heifers of a specific dairy breed.

WHAT WILL THEY COST TO BUY?

Beef calves are usually purchased at weaning (400 to 600 pounds), whereas dairy calves are purchased at a very young age; you can raise them on milk replacer until they are old enough to eat pasture, hay, and grain. Weaned beef calves cost more than young dairy calves. There are extra expenses for dairy calves, however, including milk replacer, starter pellets, and grain. A dairy heifer will be more expensive than a young male of her breed, since the dairy heifer will be worth more when she grows up. By contrast, a beef steer costs more than a beef heifer, because he will be worth more as a finished beef animal. The price of cattle fluctuates, depending on supply and demand. When prices are low, you might be able to buy a beef calf for 70 cents per pound. When prices are high, you might pay $1.40 per pound or more. So you might reasonably expect to spend anywhere from $350 to $850 for a weaned 500- to 600-pound calf. By purchasing cattle when they are high priced, you are hoping the prices stay high long enough that you will get a good price when you sell them or their offspring!

They will be more salable. A beef–dairy cross may make good beef for your freezer, a nice family milk cow, or a nurse cow to raise an extra calf, but she won't give enough milk to be profitable in someone's dairy.

What Will It Cost to Raise Cattle?

If you are fattening a yearling beef steer on pasture and grain, or hay and grain, it will cost more than raising him on grass alone. The price of grain and hay can fluctuate, but on average the hay may cost $70 to $80 per ton. In many regions, alfalfa hay of good quality will cost more than grass hay, unless it's an area where only alfalfa hay is grown. Grain costs also vary. Rolled barley may cost $50 to $54 per ton. Other types of grain can also be used. The most economical feed (hay or grain) is a kind that is grown in your area, since it will have lower transportation costs.

The amount of feed you'll need to buy will depend on whether you are supplying all of the feed or the animals are using pasture as part of their food. To feed hay through winter, you might need one and a half tons of hay per animal through the season. If you feed hay and grain to a beef steer in a corral, with no pasture, to grow him and fatten him for butchering, you might feed as much as two tons of grain during the 10 to 12 months you are growing him.

The cheapest way to raise cattle is on grass. Most ranchers keep cattle on pasture and only feed hay when grass is not good, such as during winter when it is snow-covered, if you live in a cold climate, or during drought when pastures dry up and have little food value. Cattle also do well on cornstalks after a cornfield is harvested, or grain stubble after oats, wheat, or barley is harvested. Cattle can graze these crop-aftermath pastures to grow inexpensively, then go into feedlots for finishing on grain to grow them up faster and fatten them for butchering.

Some stockmen "finish" cattle on grass alone, and sell them as grass-fed cattle rather than grain-fed (see chapter 7). Calves raised entirely on pasture don't grow quite as fast as grain-fed calves, but usually cost less to feed. So they often make just as much or more money for the stockman.

If you live in a climate with hard winters, or have snow that covers the grass, you'll have to buy hay for winter feed and keep cattle off the best pastures until the grass grows tall enough to graze again in spring. Cattle on poorer pasture can stay on it during winter while feeding on hay. Keep them on that pasture or confine them in a corral in spring while the grass in other pastures grows. If you have enough acreage, it can be divided into several portions for rotational grazing, to let each parcel regrow after being grazed.

Pasture allows you to raise a weaned calf — or several — and sell or butcher by fall without buying hay. Or you can pasture a herd of cows and calves until the calves can be sold in the fall and you'll only need hay for the cows over winter. If you have adequate pasture, or only need to purchase a little hay to go through winter, you may choose to keep your weaned calves over winter and sell them as big yearlings the next year.

Facilities and Fences

You need good fences for pens and pastures, and a reliable source of water for the cattle. In a severe climate you need some type of windbreak or shelter for them. A young calf needs more shelter than an older calf or mature animal. A milk cow needs a run-in shed or a roof of some kind, so you can milk her in a clean, dry place if it's raining or snowing. If you have a small dairy, you will want a barn.

Pen (Corral)

Even if you raise cattle on pasture, you need a pen for the times you need to capture them — giving vaccinations, treating a sick one, or helping a cow calve if she has problems. Even with just one or two calves to raise for beef, you still need a corral. It will be the safest place to put them when you bring them home. A calf brought to a strange place may desperately try to escape and go back to his familiar herd; he may crawl through a pasture fence and be gone. A corral made of poles, boards or mesh wire is more secure because he can't slip through. Barbed wire or smooth wire is not adequate for a corral because cattle may go through it if they try hard enough or crash into it.

A holding pen for a herd of cattle should be adequate to hold the whole herd comfortably with room to move around and sort them. For a herd you need two adjacent pens — one for sorting and one for the animals you've sorted — when weaning or sorting out calves to sell. If you only have a few calves, you might get by with one pen, and it can be smaller. Calves that will be spending much of their time in the pen should be in one large enough to give them room to move around. The pen should have at least 900 to 1,000 square feet for one calf. This space can be provided by a variety of dimensions: 10 feet by 100 feet; 20 by 50 feet; 30 by 35 feet, or whatever works to fit the space you have. If you have several calves, add at least 300 more square feet

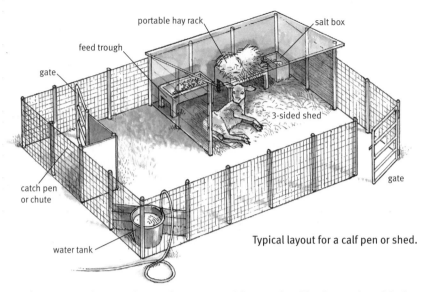

portable hay rack
salt box
feed trough
gate
3-sided shed
catch pen
or chute
gate
water tank
Typical layout for a calf pen or shed.

of space to the total area for every additional calf. There should also be shade from a building or tree. A calf needs about 100 square feet (an area 10 feet by 10 feet) of summer shade.

The pen area should be dry, with good drainage. Never build a pen on land that may flood or become swampy from irrigation water during summer. Cattle need good footing and a dry place to sleep. If necessary, sand can be put in the bedding area or along a shady spot where the cattle like to sleep, to make sure that part of the pen stays dry.

Make sure your pen and pasture have no hazards such as nails sticking out of the fencing, loose wires, or splintered boards that might injure an animal. A pole or board on the ground with nails sticking out can cause serious injury if an animal steps on it. Wire or nails lying around in pen or pasture may be eaten with feed and can puncture the animal's stomach. Don't leave baling twines hanging on a fence or lying on the ground. Cattle, especially calves, are curious and often get into trouble. They may try to eat baling twine or chew on a plastic bag that blew out of the garbage can; the indigestible material may create a blockage in the digestive tract that could kill them. If cattle are in a barn, make sure all electric wires or outlets are out of reach. Clean up old junk piles that might have hazardous items like discarded tractor or vehicle batteries, broken glass, or containers that hold chemicals, oil, or antifreeze.

Don't use electric fencing in a small pen. There might be times when you need to corner a calf to catch him for some reason, and you don't want to shock him. He might leap back from the fence and crash into you. The pen should be a safe and secure place where cattle don't have to worry about being shocked. If cattle must come into the pen from a pasture in order to be caught, you don't want them to be afraid. Once shocked, they will hesitate to go in there again. You also don't want to shock yourself. You may need to climb over the fence or might be pressed up against it when catching a calf.

BUILDING A PEN

The corral can be built of boards, poles, pipes, metal panels, or strong woven-wire netting. Make sure the fence is constructed so animals cannot jump over or crawl under or through the pen. A frantic, homesick animal in a new place may try to escape. Even if you purchase already weaned calves that aren't so desperate to get back to their mothers, a good fence is still a necessity. Someday you may be using the pen for weaning your own calves, and you want it foolproof.

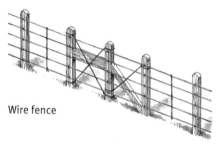

Wire fence

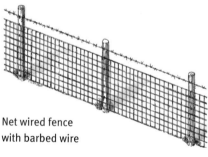

Net wired fence
with barbed wire

Use sturdy wood posts, about eight feet long — tall enough to set deeply into the ground and still make a fence at least five feet high. If you plan to pen mature cattle, make the corral fence even taller. A frightened or determined cow may be able to jump a five-foot fence; six feet tall is better.

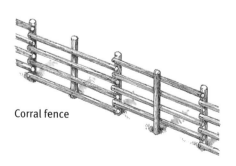

Corral fence

Secure poles, boards, woven wire netting, or metal panels to the posts, which should be set 8 to 10 feet apart for a strong fence. If posts are farther apart than that, the sections will not be as strong. Materials for a good pen will be expensive, but it will last a very long time if properly constructed.

DIGGING POSTHOLES AND SETTING POSTS

Build the pen on solid ground, preferably with some rock or gravel, or on a high spot that will never flood. Posts set in a boggy, wet area will become loose and wobbly. Postholes can be dug with just a shovel if the ground is mainly dirt with few rocks. If it's rocky, use an iron bar to jar the rocks loose as you dig.

Set the posts in a straight line. A crooked fence will not be nearly as strong. Set the corner posts first and sight between them to line up the other post holes and posts, or stretch a long string between them to give an exact line. The posts should be set at least two and a half feet into the ground, preferably three feet for a corral fence. They should be treated on the bottom end with wood preservative so they won't rot. The preservative should cover the entire portion set into the ground and extend a few inches above ground level, so that ground moisture won't rot the post.

When filling in dirt and rocks around a post, put in a little at a time, tamping firmly with an iron bar or tamping stick before adding any more dirt. The secret to having tight, solid posts is to set them deep enough and to tamp the dirt very firmly around them. They will stay more solid if you can pack mostly dirt around them rather than just rocks or gravel.

CATCH PEN OR CHUTE

You need a small catch pen or chute in one corner of your pen, where you can restrain an animal for medical purposes. For any grain-fed animal (such as a steer fattened for butcher, or a milk cow), you can make a stanchion in the feeding area, where you can lock the animal's head in place when he or she puts it through to eat. If you routinely feed the animal in the stanchion, he/she will be easy to catch.

A small chute at one corner of the pen is handy for giving injections (as you must do to vaccinate cattle against certain diseases) or to immobilize them for other procedures. You can herd the animals along the fence and into the chute, swinging the gate shut behind them. If you have a herd of cattle, you may want to invest in a factory-made metal squeeze chute that safely restrains the animal. A less expensive option is purchasing a head catcher to set at the end of a wooden chute made of poles or boards.

A swinging gate or panel will work (capturing an animal behind it) if you haven't had time to build a chute. If you don't have a sturdy gate in a location that will work, secure a stout wood or metal panel to a strong post in your corral. It should be tied at top and bottom with strong rope just loose enough so it can swing like a gate. Tie it where it will be easy to herd the animal along the fence and into the V-shaped trap made by the fence and the panel.

Before you herd the animal into this catch-trap, tie a long rope to the fence (this only works with a pole or board fence) at a position that will be just behind the animal when he goes behind the panel and has his head at the front of the V. Have the rope lying on the ground for the calf to walk over, where you can easily reach it.

It takes two people to catch a calf. One person can quietly herd

You can corner a cow behind a sturdy gate.

him along the fence and into the trap, while the other person is ready to swing the panel against the animal and pick up the rope. The panel must be swung open fairly wide so the animal will go into the trap; if the opening is too narrow, he may be suspicious.

Walk the calf over the rope, then swing the panel tight against him as you pick up the rope and loop it around a panel pole at proper height to make a barrier so he can't back out. The rope holds the calf in the V and secures the panel against him so he can't turn around.

The rope can be tied in place or held, with several loops, around the panel pole. A chute or catch-pen is the best and safest way to restrain an animal, but a panel trap can work if you don't have a chute and the cattle are not very wild. For instance, it can be a way to temporarily corner and restrain an animal if you need to put a halter on him.

Pasture

Although cattle can subsist on many types of forage plants, make sure your pasture has adequate grass or other nutritious forage and is not a weed patch. Good pasture may support two cows per acre during the growing season, whereas a pasture with sparser grass may only support one cow on ten acres. If you are unsure about the quality of a pasture or how many cattle you can support on your acreage, consult your local Cooperative Extension Service agent.

PASTURE FENCES

If there are already fences on your place, check the pasture fences carefully to make sure they are in good shape and will hold cattle. Tighten loose or sagging wires. Replace any missing staples on wood posts, or metal clips on steel posts, so the wire is securely attached to every post. Make sure all poles or boards are securely nailed to the posts and there are no "holes" or wide spots where a calf could slip through. A dry ditch or gully that goes under the fence might become an escape route where a calf might walk under the fence. Put a pole across it, under the fence, and attach it securely so cattle cannot push it away.

Repair or replace any pasture fences that are old and falling down. If there are no fences around your pasture area, build some or hire someone to build fences before you bring cattle. A pole or board fence looks nice, but is more expensive than a wire fence. Your corral should be made of sturdy poles or boards, but a large pasture area can be more economically fenced with wire.

A wire fence will hold cattle that are not being crowded against it or not trying hard to escape. The wire must be tight and not sagging. Animals that can work their heads through the fence usually try to go on through. You don't want cattle to start reaching through for grass on the other side. A tight four- or five-strand barbed-wire fence with stays (wooden or metal spacers midway between posts, to keep wires from being stretched apart) will generally hold cattle, whereas a smooth wire fence will need more strands because cattle are more apt to reach through it.

Net wire (mesh or woven wire) is best, since a cow or calf cannot crawl through it. The netting must be stretched tightly and be tall enough to keep cattle from leaning over it. One or two strands of tight barbed wire above the netting will help keep cattle from reaching over it and mashing it down.

An electric fence will keep cattle in a pasture once they learn about it. After they are shocked, cattle tend to stay away from the fence. Boundary fences around a property should be more solid, however, than an electric fence. An electric fence is merely a psychological deterrent and not a physical barrier. A calf trying to get away, or a herd of cattle running from dogs, may charge right through it.

An electric fence is most useful for keeping cattle from rubbing on, crawling through, leaning over, or reaching through a traditional fence; making it more foolproof and longer-lasting; or to divide a pasture into sections for rotational grazing. You can use portable or temporary electric fencing to let cattle graze one portion while the other parcels grow taller, moving them to a new section after they graze the one they're in. Dividing pastures can extend their usefulness, allowing the grazed portions to regrow.

For an electric fence you need a battery-powered or a plug-in fence charger, properly grounded and hooked up to the hot wire. The wire is attached to fence posts with insulators. The electric wire can't touch any metal or it will short out and won't work. It shouldn't touch wood posts or poles either, or it will short out when the wood is wet. Keep all weeds and brush clipped away from the electric fence or they also may short it out or cause a fire. No matter what kind of fence you have, check it regularly and keep it in good repair. Cattle will not keep challenging or rubbing on a fence if you augment it with an electric wire.

A simple shed provides protection from the sun in the summer and from storms in the winter.

Shelter

Most of the time cattle don't need shelter except for a windbreak (bushes and trees or a manmade barrier). Their heavy hair coat insulates them against the cold and gives some protection from wind, rain, and snow. In hot climates, however, shade is quite important. A simple roof with one or two walls can usually provide shade in summer and protection from wind and storms in winter if there are no trees in the pasture or next to the pen.

Young calves are more likely to need shelter than adult cattle. They are more susceptible to stress from cold, wet, or very hot weather. For calves, two sheets of plywood placed on each side of a fence corner can make a very adequate windbreak, and another sheet of plywood over the top will make a roof. The roof on any shed should slope so rain or melting snow will run off. Make sure it slopes away from the pen so the water won't create a mud hole in the pen or flow into the shed. The shed should be situated on a high, dry spot. A bale of straw, some bark mulch, sawdust or wood chips in the shed, or along a windbreak, will give cattle a dry place to sleep.

Before you build a shed, figure out which way the wind usually blows, and position the shed walls to give the greatest protection from the wind. If you live in an area that has hot, humid summers, protection from hot sun is also important. The roof will provide shade, but you'll also need some air flow (ventilation to create a breeze) to help keep the animals cool. In a hot climate the roof should be high and the shelter should have no walls to hinder air movement. In a cold climate the roof can be lower, with walls to help prevent heat loss on a cold day.

Water and Feed

Cattle need a continual supply of fresh water, even if it's a tank you fill with a garden hose. If you have just one or two calves and they are watered in a pen or shed, you may only need a water tub, but it should be up off the ground so they won't put their feet in it. It's hard to keep the water clean if the tub is on the ground; the calves will walk in it and soil it with manure.

Make a stand or frame to hold the tub. Keep it 20 inches off the ground but not much higher or it will be difficult for calves to drink. Nail a board across the corner of the shed or corral fence, leaving room for a tub or bucket to fit snugly. You can pull the tub or bucket up out of its holder to rinse and clean it, but the calf can't tip it over.

If you have several calves or a herd of cows to water in a corral, or a pasture where there is no stream, pond, or ditch, buy a stock tank. A simple tank made of aluminum or galvanized metal is not very expensive. An automatic waterer is easier than filling the tank with a hose, but you will have to trench a water line to it. Make sure the water pipe is deep enough below the frost line in winter and insulate the upright riser pipe that brings water up to the tank or drinking bowl. Insulated waterers or heated waterers can be used in a cold climate. Choose a type of tank or waterer that is easy to clean.

If you water cattle with a hose, drain it thoroughly after each use in cold weather so it won't freeze. Make sure the outdoor hydrant for the hose is the frost-free kind in which the water drains back down out of the upright pipe after you turn it off, so it won't freeze.

Cattle drink more in hot weather than in cold weather. Cows nursing calves drink more water because they need extra fluid to produce milk. Make sure cattle always have water and be sure that it doesn't freeze in winter. If you use a tub or simple metal tank, you may have to break ice morning and evening. If you are only watering one or two calves, a rubber tub is handy in winter because it can be tipped over

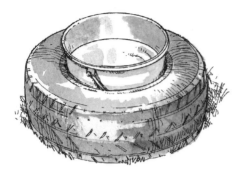

An old tire can be a good bucket holder, preventing the water from spilling.

and whacked to remove ice; it won't break or crack. Rubber can withstand more pounding than metal or plastic without developing a leak.

If you are watering calves in a corral, situate the water away from the feed rack, grain trough, or feeding area so cattle won't drag feed into the water. You also want their water far away from where they sleep. Cattle that have to walk some distance to water aren't as likely to stand close to it and soil it. Keep the water fresh and clean, even if you have to dump and rinse a tub or tank every day. If it is not clean, they won't drink enough water because they won't like the taste. Use a tub or tank that is easy to dump and rinse.

FEEDING HAY

Cattle will waste hay fed on the ground unless you place it on well-sodded ground. Cattle fed hay on bare ground or mud won't eat all of it because they don't want to clean up the part that gets dirty. They also won't eat hay that has been stepped on, laid on, or soiled with manure. Cattle are fussier than horses and waste more hay unless it is fed in a clean place.

Cattle in a corral won't waste as much if you use a feed rack or hay manger. Clean it out if they don't eat all the hay in the bottom or corners. Hay will mold if it gets wet. If you are feeding just one or two calves and they have access to a shed, you can feed them inside the shed during wet weather.

If you feed outdoors, make a little roof over the feed manger, hay rack, or grain box, so moisture from rain or snow won't ruin the feed. Moldy hay is dusty; mold spores fly into the air when cattle sling the hay around, which may irritate their air passages or cause coughing.

FEEDING GRAIN

If you are fattening a calf on grain, you'll be feeding him grain every day, splitting the ration into two portions to feed morning and evening so he won't eat too much all at once, which could lead to digestive problems, bloat, or founder. You'll need a good trough or grain box, or a rubber tub on a feed stand where he can't pull it off or step in it. Calves will not eat dirty grain.

A simple rubber tub works fine if you only have one calf, although calves do better in pairs; they eat more aggressively and gain faster. A large rubber tub will work for two calves if they get along fairly well and don't fight over the grain. The tub is easy to clean and can be rinsed out if it gets dirty. Put a roof over the tub or feed trough, since calves won't eat wet grain; moldy grain may make cattle sick. Clean out any leftover kernels of grain before adding new feed. Don't feed calves more than they will clean up within a few minutes.

An inexpensive feed trough can be made using two-inch lumber cut into lengths. If several calves will be using it, allow three square feet per calf. Make the sides of the trough at least six inches deep. Put it up off the ground so calves won't walk in it; a good height is 18 to 20 inches.

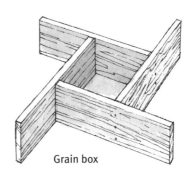

Grain box

Feed trough made from rough lumber.

The grain should always be fresh. If there are bird droppings in the tub or trough, or any buildup of old, fermented, or moldy grain in the corners, calves may refuse to eat the next batch. If it smells different or tastes bad, they won't eat it.

Manure Disposal

Cattle create a lot of manure. Out on a large pasture this is no problem, and is quite beneficial since it serves as fertilizer for plants. In a pen or shed, however, manure will become too concentrated. After a while, it builds up and must be removed. A corral is easiest to clean with a tractor and blade or loader. If you don't have one, borrow or rent one from a neighbor or hire someone to do it. The manure can be piled up for later use on your pastures or dumped into a manure spreader and hauled out for immediate spreading.

If your shed is only three-sided and tall enough, or your barn has a high roof and a large door, you can clean it out with a tractor. Otherwise you must keep it clean with a wheelbarrow and manure fork. A manure fork is similar to a pitchfork, but has more tines. Manure and straw can't fall through it so easily. If cattle spend much time in a barn or shed, clean out all the manure and soiled bedding regularly to prevent buildup.

FERTILIZE WITH MANURE

Manure makes excellent fertilizer. It contains just the right ingredients needed by plants for healthy growth. Spread the manure over your pasture or use it for the family garden. You can make a compost pile with soiled bedding from a barn or shed to enable microbes and fermentation to break down the fibrous material and create a more useful fertilizer. People in your neighborhood who have gardens may want to buy manure from you. This is often a good way to be rid of your compost pile or corral cleanings — if you don't have enough land to use it all yourself — and make a little extra money as well.

Halters and Ropes

A good halter and rope are essential equipment for cattle control, in case you have to tie one to give veterinary assistance, help a cow calve or nurse her calf, or immobilize the animal's head while restrained in a chute. An adjustable rope halter can be made from a 12-foot length of rope, or you can purchase a cow-size or calf-size halter.

When putting the halter on the animal, always place the adjustable part on his left side. To put on an adjustable rope halter, first enlarge it so it can be slipped on easily, then tighten it up (making it smaller) to be snug and close fitting, so the animal cannot pull it off or rub it off. A halter with a long lead rope allows enough length of rope to safely wrap around a corral post and still keep out of the animal's way.

TYING KNOTS

If you ever have to tie up a calf or cow, you should know how to tie a good knot. You don't want the animal to get away when tied. Learn how to tie knots that will stay tied, yet can be untied easily to release the animal, even if the cow or calf has pulled hard on the rope. Sometimes you need a slipknot that will always stay tight, and sometimes you need a knot that definitely will not slip and tighten. For your own safety and the safety of the animal, it helps to know how to tie the right kind of knot for each situation. Remember, when tying a knot, the end of the rope you are using is the working end. The rest of the rope is the standing part.

The overhand knot is the easiest knot to tie. This is the simple knot you make first when tying your shoes. This knot is often the first step in forming more complex knots.

Overhand knot

The square knot is a more useful version of the overhand knot. It is just two overhand knots, one tied on top of the other. If tied correctly, this is a perfect knot for joining two pieces of rope, such as tying a broken rope back together, or for tying a rope around a gate and gatepost to keep the gate closed. A properly tied square knot will not slip.

SQUARE KNOT

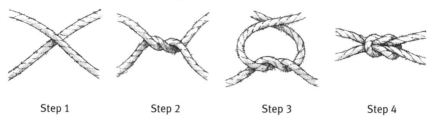

| Step 1 | Step 2 | Step 3 | Step 4 |

QUICK-RELEASE KNOT

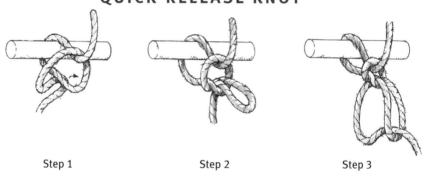

| Step 1 | Step 2 | Step 3 |

To tie a square knot, start by tying an overhand knot. Then tie another one on top of it, but this time in reverse. Before it is pulled tight, the square knot looks like two closed loops leading in opposite directions and linked together.

The quick-release knot is also called a reefer's knot, a bowknot or a manger tie. It is useful for tying an animal to a fence post or to a pole. Like a square knot, this is a non-slip knot. The quick-release knot has the advantage, however, of being more easily untied when it has been pulled really tight — after the animal has pulled back on his halter.

The bowline knot is probably the most useful non-slip knot when working with livestock. You can tie a rope around an animal's neck or body without danger of it tightening up when the rope is pulled. This is the knot to use if you must tie an animal by the neck, since it will not slip and choke the animal. It is relatively easy to untie.

BOWLINE KNOT

| Step 1 | Step 2 | Step 3 | Step 4 |

DOUBLE HALF-HITCH

| Step 1 | Step 2 |

For an easy way to remember how to tie a bowline, think of this story: The first loop is a rabbit hole, the standing part of the rope is a tree, and the working end of the rope is a rabbit. The rabbit comes out of the hole, runs around the tree, and goes back down its hole.

A double half-hitch is a quick and easy knot to tie and acts as a slip-knot, staying tight when it's pulled on. It's a handy way to secure a rope around an animal's leg when tying a leg back to keep the animal from kicking or to secure the end of the rope when no other knot seems appropriate.

To tie a double half-hitch, position the standing part of the rope to your left and take the working end in your right hand. Pass the running end of the rope over or around a post or the cow's leg. Bring the running end over the standing part of the rope, under it, then insert it into the loop (the loop around the post or leg) from the bottom. Repeat this same step to form the second half of the hitch.

CHAPTER TWO

Nutrition

To feed cattle properly, you must understand how they digest food. Cattle are ruminants. They have four stomach compartments and chew their cud — regurgitating feed to chew it again more thoroughly. The four compartments of a ruminant's stomach are the rumen (paunch), reticulum (honeycomb compartment), the omasum (also called manyplies because of its numerous plies or folds), and the abomasum (true stomach), which is similar to our own stomach. The rumen is the largest compartment.

Ruminants such as cattle, sheep, goats, and deer have a larger digestive system than single-stomached animals like humans. Cattle break down the fiber in the rumen, which is like a large fermentation vat.

When cattle eat, they chew food hurriedly — just enough to moisten for swallowing. The food then travels to the rumen, where it is softened by digestive juices and broken down by rumen bacteria. Cattle regurgitate a mass of food and liquid from the rumen and reticulum, swallow the liquid, and then chew the mass thoroughly before swallowing it again. This is called rumination. The re-chewed food passes into the omasum, where the liquid is squeezed out, then into the abomasum, and from there to the intestine.

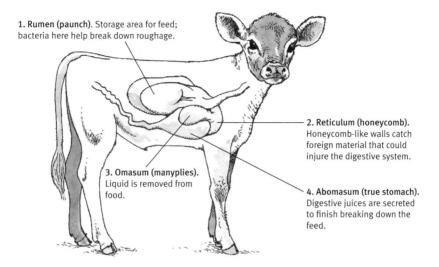

1. **Rumen (paunch).** Storage area for feed; bacteria here help break down roughage.

2. **Reticulum (honeycomb).** Honeycomb-like walls catch foreign material that could injure the digestive system.

3. **Omasum (manyplies).** Liquid is removed from food.

4. **Abomasum (true stomach).** Digestive juices are secreted to finish breaking down the feed.

Ruminants, such as cattle, have four-part stomachs.

What Do Cattle Need?

Cattle can do well on a wide variety of feeds. What you feed depends on whether you are feeding beef calves, dairy calves, or mature cattle. Specific feeding guidelines are described in later chapters. The elements of good nutrition, however, are the same for all cattle. Make sure feed contains the basic ingredients for good nutrition: protein, carbohydrates, fats, vitamins, and minerals.

Cattle can eat many kinds of feed. Feedlots can finish calves on everything from field corn to waste materials from food processing (potato skins, beet pulp, citrus pulp — whatever is cheap and handy). Cattle can make use of by-products that would otherwise be wasted or create a huge disposal problem.

Protein

Protein is necessary for growth, reproduction, and milk production. Protein can be supplied by high-quality legume hay such as alfalfa or clover, green pasture grasses, or high-quality grass hay. Protein supplements (feeds containing high levels of protein) include cottonseed meal, soybean meal, and linseed meal. Cattle on good hay or pasture don't need supplements. High-quality alfalfa is a very good source of protein, but care must be taken to avoid bloat (see page 50).

Carbohydrates and Fats

These essentials provide energy for body maintenance and weight gain. Some feeds that contain a lot of simple carbohydrates are barley, wheat, corn, milo (grain sorghum), oats, and food by-products such as millrun and molasses. Fats contain even more calories than carbohydrates. Forages contain very little fat, but there is some fat in grains. Forages provide adequate carbohydrates for most cattle; cellulose, one of the main types of fiber in plants, is a complex carbohydrate.

Vitamins and Minerals

Vitamins are necessary for health and growth. Green grass, alfalfa hay, and good grass hay contain carotene, which the body converts into vitamin A. Overly mature, dry hay may be deficient in this nutrient. The other vitamins cattle need are found in the feed or created in the gut. The exception is vitamin D, which is formed by action of sunlight on the skin. Cattle receive plenty of vitamin D unless they spend all their time in a barn.

Minerals occur naturally in roughages and grain. Most cattle do not need mineral supplements (except salt) beyond those found in ordinary feeds, unless soils in your region are deficient in certain minerals such as copper, iodine, or selenium.

Be careful with supplements. Check with your veterinarian or county Extension agent before you add mineral supplements to cattle diets to know if these are needed in your area. Some minerals are harmful if overfed. Iodine and selenium, for instance, have a very narrow margin for safety. Mineral-deficient diets are unhealthy for cattle, but excessive mineral levels can be toxic.

Calcium and phosphorus are the minerals cattle need in largest quantity. These are necessary for growth and for strong bones. Calcium is well supplied by alfalfa or clover and many pasture grasses. Grains are a good source of phosphorus, but it is also present in adequate quantities in most pasture grasses.

Salt, a combination of sodium and chloride, is very important for cattle and should always be supplied. This is often the only mineral cattle need to have added to their diet, since there is usually not enough in forages. Salt is necessary for proper body function and helps stimulate the appetite. Always provide salt, either as a block or in loose form in a salt box.

Trace mineral salt can be used if feeds are deficient. Trace minerals include copper, iron, iodine, cobalt, manganese, selenium, and zinc. Your veterinarian or Extension agent can help you figure out what kind of salt to use and whether it should include trace minerals.

Water

Water is essential and should never be overlooked. Cattle can go without feed much longer than they can go without water. Make sure your cattle have a constant supply of clean water. They will not eat enough feed if they are short on water.

The larger the animal, the more water it needs. A 350-pound calf will drink 1 to 5 gallons per day. A 500-pound calf needs 2 to 6 gallons, a 750-pound animal needs 10 to 15 gallons, and a 1,000-pound steer needs more than 20 gallons per day. A cow that is nursing a calf may drink 30 or more gallons. All cattle need more water during hot weather than in cold weather.

A stream or river running along or through your place may always have water, but if cattle must drink from a ditch or pond, make sure it always has water in it. If the ditch dries up or gets turned off or diverted somewhere else before it gets to your pasture, the cattle may be without water. A trough or automatic waterer is often a more dependable and safer source of water.

Check the water source often, especially if it freezes over in winter and you must break ice. Make sure cattle can access the water easily. If they must walk on ice to reach a chopped-out hole, they may not do it unless you put sand on the ice so it won't be slippery. When there is too much ice on a pond or creek, water cattle in a tank, and keep the ice chopped out of it or use a tank heater. Make sure automatic waterers are working properly.

Roughages and Grains

Roughages (forages) are the most natural feeds for cattle, but the animals don't grow as fast or fatten as quickly on a forage-only diet as they do when grain is added to their ration. If you raise beef animals and are not trying to push them for fast weight gain, or weaned heifers to keep as cows, you should feed forages and little or no grain. A grain-fed heifer may become too fat, which can hinder her reproduction abilities and milk production. Grain has more calories per pound of feed than hay or pasture, and is often fed to finish beef calves, enabling them to reach butcher weight and condition more quickly. Their diet can be a combination of grain and forage.

Roughages are feeds such as hay, pasture, silage, or beet pulp that are high in fiber and low in energy. Cattle with access to good green pasture have all the nutrition they need. If cattle are not on pasture, you'll have to figure out how much hay or alternative types of roughage to give them.

In cold weather, cattle need more feed to generate body heat to keep warm. Roughages provide more heat energy than grain due to the breakdown of fiber into useable energy. When weather is cold, increase their ration of grass hay.

A weaned calf needs about three pounds of hay per day for each 100 pounds of body weight. For example, a 500-pound calf needs about 15 pounds of good hay. The actual amount may vary, depending on the type of hay being fed and its quality. A good rule of thumb is to provide all the good hay that calves will clean up. If they start to waste a lot, you are feeding too much, unless they are not cleaning it up because it is coarse, stemmy, and of poor quality. Cows will eat poorer quality or coarser hay than calves will; calves must have very fine, palatable hay.

Alfalfa hay is richer in protein, vitamin A, and calcium than grass hay, but can also cause digestive problems and bloat, which can be deadly. Alfalfa tends to ferment too quickly in the rumen, creating gas. If the rumen fills with gas, it becomes large and tight like a balloon, putting pressure on the calf's lungs and suffocating him.

First-cutting alfalfa can be ideal feed for cattle; it usually has a little grass in it. Later cuttings of alfalfa are often richer and more apt to cause bloat. It's always safer to feed a mix of grass and alfalfa. Alfalfa molds more readily than grass hay if it gets wet or was baled too green (before it was completely dry).

When buying hay, check to make sure it doesn't contain mold. Open a few bales to look at the hay inside. Beware of any bales that seem heavier than average; they may contain moisture and mold. Bales from the top or bottom of an uncovered stack are apt to be moldy, due to moisture from rain or snow or absorbed from the ground.

Hay quality

Make sure that hay you buy or harvest yourself is not moldy or dusty, or it may give cattle digestive or respiratory problems. Many types of mold are relatively harmless, but some are quite toxic and can poison an animal or cause abortion in pregnant cows. Dusty hay can irritate air passages, which can open the way for pathogenic organisms, causing pneumonia.

Overly mature hay is stemmy and coarse and cattle will waste it. The protein and nutrients in forage plants are mainly in the leaves or grass blades. If hay is mostly stems, it is predominantly fiber and does not contain as many nutrients. Stemmy hay is also difficult to chew.

When buying alfalfa, make sure it is bright green (not yellow or brown) with lots of leaves and small stems. Alfalfa hay that is cut early, just before it blooms, is finer and more nutritious. After it blooms, it has less protein. It has larger stems and less leaves. Grass hay should also be bright green (not brown and dry), and cut while still growing and full of nutrients. Any hay that is overly mature when cut has lost much of its food value. If you don't have much experience with hay, ask your Extension agent or another knowledgeable person to help you select hay to buy.

Grains

Concentrates are feeds that are more nutrient-dense than grass and hay; they contain more calories per pound. Grains like corn, milo, oats, barley, and wheat are concentrates that are sometimes fed to cattle when pastures are poor, hay is in short supply, or to finish beef calves faster. In the Pacific Northwest, barley is plentiful and less expensive than milo or corn. Wheat is usually too high-priced to feed to cattle. Corn is high in energy and commonly used for cattle in regions where it is grown. Oats also make good feed, as does dried beet pulp moistened with molasses.

Cattle eat steam-rolled or crimped grain more easily than whole grains, utilizing more of its food value. Don't use grain that is too finely-ground or it may cause bloat or diarrhea and cattle may also be reluctant to eat it. Your Extension agent can advise you on what types of grain to use if you are fattening cattle on grain.

Pasture Management

Pasture should contain several types of forage plants that are nutritious for cattle. Most pastures are a mix of grasses, and some contain a legume as well (clover, alfalfa, or birdsfoot trefoil). A mixed pasture is ideal for cattle because the legumes have a higher protein level than most grasses and may stay greener after some of the pasture grasses mature and lose nutrient quality.

Pasture Care

Some pasture plants become coarse when mature, and cattle won't eat them. Weedy areas can also be a problem. You can improve the pasture if you mow or clip weeds so they won't go to seed and spread. If certain parts of the pasture grow too tall and coarse, cattle leave those and keep eating the shorter, younger grasses. Entice cattle back to these areas by clipping the tall plants to allow regrowth that is more palatable and nutritious. Seed bare spots with a "pasture mix" of grasses, scattering seeds by hand when the ground is wet.

Pasture in rainy areas will grow nicely without any help, but pasture in a dry climate will dry out and lose its food value by late summer, unless you keep it watered with a ditch or sprinklers. Learn how to water the pasture to keep it green.

Lush green grass has about as much protein and vitamin A as good alfalfa hay. For growing calves or cows with calves, good pasture is excellent feed. Keep a close eye on the grass. Don't just turn cattle out and forget about them. If pasture plants become short or too dry, cattle won't do as well. Dry plants are not very nutritious; they lose their protein and vitamin A. Under these conditions, feed good hay to help supplement the pasture.

Check your pasture for poisonous plants. Some kinds grow in wet areas or along a ditch or stream. Learn how to identify them. The Cooperative Extension Service should have a book with illustrations to show what the plants look like. If you find a strange plant in your pasture, take a sample to your Extension agent to find out what it is, or consult your veterinarian. Cattle can be poisoned by larkspur, hemlock, and many other toxic plants; if you have any of those in a pasture, get rid of them. Different geographic regions have different species of poisonous plants. There are hundreds of plants nationwide that can be toxic. Find out which ones might be a problem in your region.

ROTATIONAL GRAZING

Even if you have only a small acreage, grass will last longer if you divide pastures into two or three pieces, grazing them alternately. Cattle that stay full time in the same area will overgraze their favorite grasses and leave others to grow tall and coarse and go to waste unless there is nothing left to eat. This can be harmful to a pasture over time. Plants are not healthy if overgrazed or undergrazed.

Divide the pasture and put cattle into the first segment when grass is at least four to six inches tall. Move them before they graze it too short; the grazed pasture will regrow and can be used again. Rotational grazing can extend pasture by keeping more of it in the four-to-six-inch phase, which cattle like and is easiest for them to eat, and less of it grazed down to the ground or going to waste because it got too tall and coarse. Move them before grass gets too short; it won't take so long to regrow and you can come back to it more quickly. Divide the pasture with permanent fence or with moveable electric fencing.

CHAPTER THREE

Maintaining Healthy Cattle

THIS CHAPTER DESCRIBES WAYS to prevent illness in cattle. It also explains how to recognize illnesses and what to do when an animal gets sick. Some illnesses are not serious, but others could kill the animal unless you provide medical treatment. Humans are vaccinated against diseases like tetanus, mumps, measles, and polio. Cattle are vaccinated against things like tetanus, malignant edema, blackleg, brucellosis, leptospirosis, and respiratory illnesses.

There are diseases for which there are no vaccines, such as bloat and some types of diarrhea. Most illnesses can be prevented with good care, proper nutrition, maintaining a clean environment, and limiting exposure to contagious diseases that are transmitted from one animal to another.

Talk to your veterinarian about a vaccination program to protect cattle against diseases in your area. A single calf or a small herd may not be exposed to diseases that might affect cattle in larger groups, but some pathogens live in the soil (blackleg, malignant edema) and your animal may be exposed even if he never sees another bovine. Other diseases, like leptospirosis or brucellosis, can be spread by wildlife. The only way to protect cattle from these diseases is to vaccinate.

Signs of Sickness

Prompt treatment of a sick animal can often make the difference between life or death in some types of illness, or can help an animal recover much more quickly. It is important to check cattle often. If you get to know your animals well and pay close attention to them every day, you'll learn how to tell whether they are feeling fine or becoming sick.

Behavior and Appearance

A healthy animal is bright and alert and has a good appetite. He comes eagerly to feed (if you are feeding hay or grain) or grazes in the pasture during times of day the cattle normally graze. A sick animal may spend most of his time lying down instead of grazing. He may be dull, with ears drooping instead of up and alert. He may stop chewing his cud because of pain, fever, or a digestive problem. Any animal that is dull and droopy and not chewing cud is probably sick. Check closely any animal that is off by himself when others are grazing. A sick animal often leaves the herd.

An animal that feels good usually stretches when he first gets up from a nap, and has an interest in his surroundings, responding with curiosity to sounds and movement. The healthy animal grooms himself. When he walks, he moves freely and easily.

By contrast, a sick animal may be less interested in things around him, concentrating on his own pain. When he does get up, it may be slowly or with effort. All his movements are slow. A sick animal lacks the sparkle of vitality and health shown by a normal cow or calf. The more serious the illness, the more indifferent the animal is to his surroundings, and the more reluctant to move.

If the animal is overly alert or anxious, however, and constantly looking around, this can also be a sign of pain, irritation, or discomfort. Some kinds of pain, such as abdominal pain, may make the animal restless — wandering around, lying down repeatedly, kicking at its belly or switching its tail, or looking around at its belly. Diseases that effect the nervous system may also make an animal hyperactive.

Respiration Rate

Breathing rate can give a clue about sickness or health. A sick animal with a fever may breathe faster than normal. Remember, though, that exercise or hot weather will also make him breathe faster. Cattle don't have many sweat glands and cool themselves by panting — creating more air exchange in the lungs to help dissipate body heat. When standing quietly or lying down, a calf or cow's respiration should be about 20 breaths per minute; it will be higher on a hot day.

Temperature

Normal body temperature in cattle is 101.5°F. This is an average; an animal with a temperature of 101° or 102°F is probably not sick. Anything higher than 102.5°F should arouse suspicion and you should closely check for any other signs of illness. Learn how to take temperature with a rectal thermometer. You can purchase one from your veterinarian. Tie a string to the ring end so you will never lose it in an animal's rectum.

If you don't have an animal thermometer, use a human rectal thermometer, but tape a string to the end of it. The main problem with a human thermometer is it does not go as high; a fever of 107°F (*very* serious!) is higher than the human thermometer can register. In most cases, it will be adequate to give an indication of whether or not the animal has a fever or a subnormal temperature. The latter can also be an indication of a serious condition.

To take temperature, first restrain the animal in a chute or tie him up securely so he can't get away. Shake the thermometer down below

98°F and lubricate the end of it with petroleum jelly (or even your saliva in the palm of your hand) so it will slide in easily. Gently lift the animal's tail and insert the thermometer into the rectum. Aim it slightly upward and put it in with a twirling motion; it will go in more easily and be less apt to poke the rectal wall.

Keep track of the thermometer. The animal may expel it with a glob of manure, so hold onto the string just in case. Leave the thermometer in the rectum a full two minutes or more for an accurate reading. When you take it out, wipe it with a paper towel or with a handful of clean straw so you can read it.

Other Signs of Illness

Other indications of the health of the animal are his eating habits. Does he chew and swallow properly, or is swallowing painful? Is he drooling or dribbling food, or having trouble belching and chewing his cud?

Are bowel movements and urination normal? With some digestive problems, the animal does not drink enough water and becomes constipated. Manure becomes firm and dry, or even absent if there is a gut blockage. With other types of digestive problems or infections, the animal may have diarrhea, squirting loose or watery manure that may be off-color or foul-smelling. Normal manure is moderately firm (not runny and watery, nor hard) and should be brown or green in color. Urination becomes difficult if there's a blockage in the steer's urinary tract due to a bladder stone. He may kick at his belly in pain, or look as though he were trying to stretch out his body (to move the blockage). If a steer shows any of these signs, call your veterinarian.

Pay attention to abnormal posture. An animal resting a leg and not putting weight on that foot may be lame. Arching the back ("humped up," with all four legs bunched together) is usually a sign of pain. A bloated animal may stand with front legs uphill, for easier belching of gas. A sick animal may lie with his head tucked around toward his flank. If an animal does not want to rise, it may mean he doesn't feel well.

Common Health Problems

Good care can help prevent many problems. A good management program, such as calfhood vaccination, annual vaccinations for the cattle herd, and proper sanitation (keeping animals in clean conditions rather than crowded and dirty) can reduce the incidence of disease. If you have a spare pen, isolate newly purchased animals for three weeks to make sure they are not coming down with a disease, and give them their vaccination during this time. Some types of disease are most common in young calves; older animals develop immunities and rarely develop those problems.

Scours

The most common problem in young calves is infectious diarrhea. Scours, caused by a variety of pathogens, is another term for diarrhea. The manure becomes runny or watery, and may be foul smelling and off-color. Scours is usually not a problem in older, weaned calves, but can be life-threatening in a young calf. Viral scours tends to affect calves in the first two weeks of life, whereas bacterial scours can strike calves up to two or three months old. Protozoal diseases like coccidiosis also can cause diarrhea. Young calves are adversely affected by scours because severe dehydration can be deadly. Preventing and treating calfhood diarrhea is covered in chapter 10.

Pneumonia

After scours, the most common killer of young calves is pneumonia. It can also strike older animals if they are severely stressed. Pneumonia, an infection of the lungs, is caused by viruses or bacteria. Germs that cause this infection are usually lurking in the environment and only make an animal sick if immunity is poor and resistance to disease is lowered by stress, as from a long trip in a truck without feed or water or exposure to wet and chilly weather. Calves weaned during bad weather suffer more stress and are susceptible to pneumonia. Cattle hauled a long way to a new home or to a sale are at risk, especially if they come in contact with cattle in a sale barn that may be sick.

A calf that doesn't nurse soon enough after birth, or does not receive enough colostrum (first milk from the dam, see chapter 9) won't receive antibodies he needs for immunity against common pathogens that cause pneumonia. A young calf weakened and stressed by scours may develop pneumonia. Calves between the ages of two weeks and two months are very susceptible to pneumonia. At this age, temporary immunity from colostrum antibodies is declining and young calves have not yet developed their own immunity.

Poor ventilation in barns or sheds can lead to pneumonia due to irritating effects on the lungs from dusty bedding and ammonia from urine and manure. Other stressful conditions include overcrowding (too many animals in a small area) or sudden changes in weather from one extreme to another.

Symptoms/effects: Pneumonia can be swift and deadly, or mild. If you see warning signs early and treat immediately, pneumonia is easier to clear up. A cow or calf coming down with pneumonia usually quits eating, lies around a lot, or stands humped up because of pain in its chest, looking depressed and dull. Ears may droop. Respiration may be fast or labored and grunting. The animal may cough or have a snotty or runny nose.

Prevention/treatment: Confine the animal and take its temperature. If it is higher than 103°F, the animal is sick and should be treated. A temperature of 104°F or higher is serious. An appropriate antibiotic should be given immediately, even if the pneumonia is caused by a virus. Antibiotics combat secondary bacterial invaders, which are deadly killers. If you are inexperienced in treating cattle, call your veterinarian to come look at the calf and prescribe an appropriate antibiotic. The veterinarian can help with treatment and leave medication with you, along with instructions for continued treatment. Even if a long-lasting (sustained release) antibiotic is given, a serious case may need one more follow-up treatment three days later.

An animal with pneumonia needs good supportive care. This means shelter to keep warm and dry and out of bad weather, and plenty of fluids. Your veterinarian can prescribe medication (usually an injection of an anti-inflammatory drug) to help reduce pain and fever so the animal might feel better and start eating and drinking again. If you can not immediately obtain the prescribed, injectible anti-inflammatory drug, you can reduce pain and fever in a calf by dissolving two aspirins in a little warm water and squirting the mixture into the back of the calf's mouth with a syringe. Fever can cause dehydration. If a young calf is not nursing, give fluids by stomach tube or esophageal feeder.

Do not stop antibiotic treatment too soon. The animal may be feeling better and eating again, and fever may be down; you might be tempted to halt the medication. But if symptoms return, the animal may be twice as hard to save. If giving a daily dose, continue with antibiotics for a full two days after all symptoms are gone and the temperature is normal again. If using a long-acting antibiotic (every three days), give a second dose if the animal is not completely normal by day three. If he does not show complete recovery within a day or two after the second dose, consult your veterinarian; the animal may need a change in antibiotic. It's not unusual to have to treat a serious pneumonia for a full week or longer.

"Red Nose" and Other Viral Respiratory Diseases

Calves should be vaccinated against virus-caused respiratory diseases such as "red nose" (IBR, infectious bovine rhinotracheitis), BVD (bovine viral diarrhea, a viral disease that can also cause respiratory disease and compromises the immune system to make cattle more susceptible to other diseases like IBR and pneumonia), PI3 (parainfluenza type 3) and BRSV (bovine respiratory syncytial virus). These viral diseases weaken the calf's immune system; bacteria move into the lungs and cause pneumonia.

Red nose (IBR) is caused by a herpes virus that can produce respiratory disease, abortion in pregnant cows, or eye problems that look like pinkeye. It is one of the most common viral infections in cattle and can spread rapidly if cattle are confined in groups in small areas.

Symptoms/effects: Signs of the respiratory form of IBR include high fever (104 to 107°F), red nose (inflammation of nostrils and muzzle), loss of appetite, rapid or difficult breathing, dullness, and profuse nasal discharge — clear and watery at first and then sticky and yellow — hanging from the nose in long strings. Watery discharge from the eyes becomes sticky as the eyelids become inflamed. All animals in a group may become infected, and many will cough.

Prevention/treatment: IBR can be prevented by vaccination. All cattle should be vaccinated annually, or even twice a year, if your veterinarian recommends it. The best immunity is given by a modified live-virus vaccine, but this type of vaccine may cause abortion in pregnant cows unless they already have immunity from previous vaccinations. A killed vaccine can be used in pregnant cows, but does not give long-lasting immunity. One way to fully protect your herd is to vaccinate annually with live-virus vaccine when animals are not pregnant (after calving and before rebreeding, for cows), and use a killed-virus vaccine in the pregnant cows six months later.

Calves should be vaccinated at six months of age. If IBR has ever been a problem in your herd, your veterinarian may suggest vaccinating calves at a young age for their first shot and revaccinating them at six months. You can use a vaccine that protects against BVD and PI3 (and sometimes some of the other viral-caused respiratory problems) at the same time. Discuss this with your veterinarian.

There is no effective treatment for IBR, BVD, and other viral respiratory problems, since antibiotics do not affect viruses. The best treatment is good supportive care. Keep the animal sheltered, eating and drinking, so its body can fight the virus. Prevention (by vaccination) is the safest route.

Diphtheria

Diphtheria is an infection in throat and mouth, often caused by the same bacteria that cause foot rot (see page 46). The bacteria enter via injury to tissues of the mouth or throat. Diphtheria can become serious if swelling in the throat obstructs the windpipe and inhibits breathing. It may also progress into the lungs and cause pneumonia. Young cattle up through two years of age seem most susceptible. Emerging teeth, after the calf sheds baby teeth, or injuries in the mouth from coarse feed or sharp seeds may open the way for infection.

Mature cattle are less affected by diphtheria. They have developed some resistance, and are less seriously affected if they do become ill, since throat and respiratory passages are much larger. They generally are not at risk of suffocation from swelling.

Symptoms/effects: The animal with diphtheria may be dull and not interested in eating. He may slobber and drool, with swellings in cheek tissues. Breath may smell bad. Swelling at the back of the throat can block the windpipe in young animals, making breathing noisy and difficult. The animal may cough, especially after exertion, and drool saliva because of difficulty swallowing. Fever may be mild (103°F) if infection is just in the mouth, or serious (106°F or higher) if the infection is deep in the throat. A calf may die from the infection, or from obstruction of air passages (suffocation) unless treated promptly. Treating a calf with diphtheria should be done quietly and carefully as calves with decreased oxygen often become agitated and this can sometimes lead to collapse.

Here is one way to tell the difference between diphtheria and pneumonia when an animal is having trouble breathing: An animal with pneumonia (damaged, infected lungs) has trouble pushing air out; an animal with diphtheria (swelling in the throat that obstructs the windpipe) has trouble drawing in air. If you watch closely, you can tell which phase of respiration is affected.

Prevention/treatment: The best prevention for diphtheria is to keep cattle in a clean environment, rather than confined and over-crowded. Treatment includes antibiotics and anti-inflammatory medication (to reduce swelling in the tissues so the animal can breathe easier), given as soon as possible. Your veterinarian can advise you on which antibiotics and medications to give.

Protozoa-Caused Diarrhea: Coccidiosis and Cryptosporidosis

Some types of intestinal infection are caused by protozoa — tiny one-celled animals that damage the intestinal lining, causing diarrhea. Coccidiosis can be debilitating in calves three weeks old or older, up through weaning age. This disease doesn't affect really young calves;

it takes 16 to 30 days for signs of diarrhea to appear. Cryptosporidiosis causes similar illness, but appears in younger calves; they are most susceptible between 5 to 20 days of age.

Symptoms/effects: Both types of protozoa cause diarrhea, which can become quite severe. With coccidiosis, there is often blood in the loose, watery feces. Protozoa are passed in the manure of sick calves and "carrier" animals. The carriers are not sick, but they have a few protozoa living in their intestines. Calves may pick up coccidia or the cryptosporidia when consuming contaminated feed or water, or by licking themselves after lying on dirty ground or soiled bedding, or when nursing their mothers if the cows have manure on the udder from lying in dirty bedding.

Symptoms of both diseases include watery diarrhea (brown or blood-tinged with coccidiosis, green/brown or cream/tan color with cryptosporidiosis), no appetite, and weight loss. A calf may be severely dehydrated.

If a calf develops loose manure, especially with blood in it, or if he is straining a lot after passing loose bowel movements, suspect coccidiosis. Some calves strain so hard, due to irritation to the large intestine, the rectum prolapses. If this happens, have your veterinarian replace the prolapse and put a few stitches across the opening to keep the calf from prolapsing again. Stitches can be removed after recovery.

Cryptosporidiosis can infect humans and can be very serious in infants or people with stressed immune systems. Always wash your hands thoroughly after handling any scouring calf.

Prevention/treatment: Antibiotics are of no help to a calf with protozoal diarrhea; the life cycle of the pathogen has already finished by the time diarrhea begins. Good care and supportive treatment (fluids and electrolytes to combat dehydration, and medications to slow and soothe the gut) may save the calf. Other calves in the group can be protected from coccidiosis by giving them anti-protozoal medication *before* they show symptoms. Your veterinarian can advise you on a control program.

The best prevention for protozoal diseases is to keep animals in a clean environment where they are not eating contaminated feed or lying in feces from sick calves. Do not bring in any new animals that might have the disease. Cryptosporidia are often introduced to a farm when young infected calves (such as dairy calves) are purchased; the calves have the protozoa and break with diarrhea after they arrive. *Never* purchase a young calf that is ill.

Coccidiosis is widespread. Most cattle have a few protozoa in the intestine. Adult cattle have some immunity and are not seriously affected; the protozoa do not multiply rapidly in adult cattle unless their immune systems are severely compromised by stress or other diseases, but adult cattle may pass a few protozoa in their manure, thereby infecting their calves.

The biggest problem occurs when young, susceptible animals are confined (at weaning time, in feedlots, and in calving areas) and exposed to a buildup of manure in dirty pens or small pastures. If they come in contact with too many protozoa, the organisms multiply in the intestine in large numbers and create damage that results in diarrhea. If animals must be confined, supply clean bedding. Keep feed and water supplies from being contaminated with manure.

Blackleg and Other Clostridial Diseases

Blackleg is a serious disease caused by bacteria that live in the soil. The clostridia family of bacteria (which includes tetanus, malignant edema, blacks disease, redwater and enterotoxemia) multiply via spores (dormant stages) that can survive for a long time and infect animals when conditions are right. These bacteria can produce deadly toxins.

Symptoms/effects: Cattle with blackleg become sick suddenly, and most of them die. Blackleg primarily affects cattle younger than two, causing inflammation of the muscles, severe toxemia ("blood poisoning" that affects the whole body) and death. The animal becomes very dull, with a high fever. Crackling swelling, caused by gas bubbles, may be felt under the skin. The swollen leg becomes hot and painful, then cold and numb. The animal usually dies within 12 to 36 hours. Often the animal dies so quickly that you don't see him sick; you just find him dead.

BE CAREFUL WITH MEDICATIONS

When using any drug or vaccine, always read labels and follow directions for dosage, how to give it, and how often. If you do not understand any of the directions, ask your veterinarian. Store medicines properly. Some need to be refrigerated. Some need to be kept out of the sunshine. Some need to be shaken well before use. Read the label!

Cattle with redwater (so named because there is usually blood in the urine) are also affected so swiftly that you generally do not find them in time to treat them. Malignant edema is caused by bacteria in soil that enter through a deep wound or injury to the reproductive tract in a cow after a difficult calving. The animal becomes dull, with swelling around the wound, a high fever, and usually dies within 24 to 48 hours.

Prevention/treatment: There is usually not time to treat blackleg or other clostridial diseases. If the animal is still alive when found, massive doses of penicillin may sometimes save an animal. The animals are rarely found in time, however. Many cattle died of blackleg in earlier times. Blackleg was one of the first livestock diseases for which a vaccine was created. Every calf should be vaccinated between two to four months of age and given a booster at weaning time, to give life-long immunity.

By contrast, redwater requires semiannual vaccination to keep cattle protected. Where redwater is a problem (northwestern Rocky Mountain region, or the Southeast), cattle must be vaccinated every four to six months. Consult your veterinarian to learn what type of vaccination program you should use.

There is a combination vaccine to protect cattle against most of the deadly clostridial diseases, including blackleg, blacks disease, redwater, malignant edema and several strains of bacteria that cause enterotoxemia. A separate vaccine can be given to protect against tetanus, if your veterinarian recommends it.

Brucellosis

This is also called Bang's disease, or undulant fever when it occurs in humans (recurring symptoms of fatigue, fever, chills, body aches, or chronic arthritis, attacks of the nervous system or emotional disturbances). This is the most common cause of abortion in cattle worldwide, except in countries where it has been controlled by vaccination. Because of the threat to human health, there was a rigorous program to eliminate the disease in cattle, and most states are now free of it except where it exists in wildlife herds.

Symptoms/effects: Brucellosis causes abortion in cows or birth of a diseased calf. The cow may have a fever before she aborts, but may not have been obviously sick. A cow that aborts may retain the placenta (it does not shed properly and hangs out for days). It is rare today to have a cow abort from brucellosis, unless it is introduced to a herd via wildlife or an infected animal. The disease will spread through the whole herd unless the females have been vaccinated.

Testing herds and removing all infected animals, along with vaccination requirements for all heifers, has nearly eradicated the disease in cattle in the United States. In many states beef cattle no longer have to be vaccinated. Since the disease is also spread by elk and bison, some regions, such as Yellowstone Park, are still experiencing problems. Farmers and ranchers in states where brucellosis persists in wildlife must continue to vaccinate all their heifers.

Prevention/treatment: If you live in a state where Bang's is still a risk to cattle, all heifers must be vaccinated between 2 and 10 months of age. Steers and bulls do not need vaccination, since they do not spread it; it is primarily spread through contaminated discharges from an aborting cow and infected tissues of aborted fetuses.

If you buy heifers or cows in a state that has a vaccination program, they should have a small metal tag in their ears with a number on it. The number will be tattooed in the ear, along with a number that specifies the year the cow was vaccinated. Ask your veterinarian if cattle in your state must be vaccinated against brucellosis. If your state has a vaccination program, you must have your veterinarian vaccinate all heifer calves when they are young.

Leptospirosis

This disease is caused by bacteria and spread by rats, mice, and other rodents, as well as by many kinds of wildlife and infected domestic animals such as pigs, dogs, or other cattle. There are many different strains of leptospirosis. Some types cause disease primarily in rodents, other types affect dogs, while others are more common in pigs, for instance. Several types can affect cattle. Bacteria are shed in the urine of an infected animal. Cattle can get "lepto" from contaminated feed or water.

Symptoms/effects: This is usually a mild disease in cattle but can have serious side effects, such as abortion in pregnant cows (even if the cow did seem sick). A few animals show fever, loss of appetite, reduced milk production, or difficult breathing. On rare occasion the sick animal may die. Young cattle are usually more severely affected than adults. If an animal has lepto, antibiotics may help. Consult your veterinarian for diagnosis and treatment.

Prevention/treatment: Vaccinate all female cattle against leptospirosis. If more than one type is causing problems in your region, use a vaccine that protects against several strains. Heifers and cows should be vaccinated at least once a year, and many veterinarians recommend vaccination every six months.

Lump jaw

Lump jaw is a swelling that occurs when cattle develop an abscess from infection in the mouth tissues. The usual cause is a puncture that opens the way for bacteria. The animal may have eaten hay or grass with sharp seeds that poked the mouth. Foxtail or downy brome ("cheat grass") seed heads have sharp stickers that may catch in the mouth.

Symptoms/effects: The abscess (pus-filled lump) may swell as large as a tennis ball or bigger, and the animal's cheek

The abscess swelling should be lanced at its lowest point.

bulges outward. The lump is often on the side of the face, or toward the back or bottom of the jaw.

Prevention/treatment: Best prevention is to avoid feeding cattle hay that contains cheat grass or foxtail (or other types of plants with stickery seed heads), and to remove any patches of this grass in a pasture. These are not very productive grasses and are not worth the problems they cause.

An abscess may eventually break and drain on its own, but the infection can be cleared up faster and with less scar tissue if lanced and drained. If you don't want to try this yourself, ask your veterinarian or an experienced person to do it. The abscess can be drained once it has come to a head, with a pus pocket that can be flushed out. Using a scalpel or very sharp knife, lance it at the lowest point of its soft spot. Pus can then be squeezed out and any remaining pus flushed out with an antiseptic solution.

Another type of lump jaw is more serious; the infection is in the bone rather than soft tissues. Bacteria enter the mouth the same way, but the puncture penetrates deeper — into the bone. Infected bone creates a painless, hard enlargement. The lump on the jawbone is immobile. Here is one way to tell the difference between a soft-tissue abscess (that can be lanced and drained) and a bony lump: The latter cannot be moved around with your fingers — it is part of the bone. The infection may eventually break through the skin, but lancing is not helpful because the bone infection will not drain. A bony lump can be treated with an intravenous injection given by your veterinarian, but the treatment is not always successful.

Foot Rot

This infection is caused by bacteria that live in soil (the same bacteria cause navel infection in newborn calves and diphtheria). Cattle may develop foot rot if there is a break in the skin of a foot. Wet areas are likely places to pick up the bacteria.

Symptoms/effects: Foot rot causes swelling, heat, and pain in the foot, resulting in severe lameness. The swelling and lameness develop very suddenly. The animal may be fine one day and severely lame the next. The foot may be too sore to put weight on it. The swelling may be at the back of the foot, or in the front between the toes. After a day or two the swelling may break and drain.

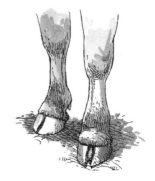

Prevention/treatment: The best prevention is to keep cattle out of boggy areas and give them a dry place in winter. They are more prone to foot rot when skin is soft from walking in water or mud; the soft skin is more easily scraped or punctured. Cattle that walk in

Foot rot causes swelling between the toes.

sharp rocks, stubble or on rough frozen ground, may also get nicks and scrapes that open the way for infection.

Animals that develop foot rot should be treated immediately and isolated from the herd. If the infection is cleared up quickly the animal won't be spreading bacteria around your farm.

Foot rot heals quickly if treated early, during the first day or two of lameness. Injection of an appropriate antibiotic (giving antibiotic coverage for at least three days) usually stops the infection. If the animal has been lame for several days and there is more damage to the foot, you may also give sulfa boluses by mouth. The two antibiotics working together can clear it up faster. Your veterinarian can recommend a course of treatment.

Pinkeye

This is usually a summer problem; bacteria that cause it are spread by face flies. Pinkeye occurs when dust, flies, bright sunlight, or tall grasses irritate the eyes of cattle. Flies carry bacteria from one animal to another. Pinkeye may not be a problem if you have only a few cattle, unless cattle on your neighbors' place are within flying distance for face flies.

Symptoms/effects: The animal holds the affected eye shut because it is very sensitive to light. The eye waters profusely. After a day or two, a white spot may appear on the cornea (the front of the eyeball). If it gets worse, the spot gets larger and the cornea becomes cloudy and blue. A severe case, if neglected, may eventually rupture the eyeball.

Prevention/treatment: Pinkeye always causes pain, and a serious case of pinkeye may cause blindness. It's better to treat the eye and make sure it heals quickly. An injection of the appropriate antibiotic can help the animal fight off the infection, and a topical antibiotic (put directly into the eye) can help.

If you use a topical medication (antibiotic spray or powder to put into the eye), the animal's tears soon wash it out, so you must repeat the treatment twice a day to be effective. Restrain the animal in a chute or tie it up so you can come close enough to squirt or puff medication into the eye. A mild case may clear up within a couple of days, but a severe case may take much longer.

If you don't want the chore and challenge of treating the eye twice daily, use longer-lasting medication, such as an injection of long-lasting antibiotic. Your veterinarian can inject a mixture of antibiotic and anti-inflammatory drugs (to reduce pain and swelling) directly into the inside of the eyelid, where it will have a much longer effect. This will clear up a severe case.

If the eye is already ulcerated and damaged, it can be treated with an eyelid injection and protected from bright sun and further injury by using a glue-on eye patch to cover it, or by your veterinarian stitching the eyelids together. Keeping the eye closed and covered for a couple weeks allows it to start healing. Even a blind eye can be healed in this fashion; the animal will regain sight, if the eyeball has not yet ruptured. Don't wait and see whether the eye will clear up on its own. Early detection and treatment is always preferable to waiting until the eye is severely damaged.

The best way to prevent pinkeye is to control face flies with insecticide (ear tags work best) and reduce the irritants that leave eyes vulnerable to infection (dust and tall grasses that brush the eyes). Clipping pastures that get overgrown can also help. There is a vaccine for pinkeye, but it is not always effective.

Cancer Eye

Ocular squamous cell carcinoma is the most common type of cancer in cattle. White-faced or light-skinned animals are most susceptible, but it can occur in dark-skinned animals. It is most common in older animals. There are two types of cancer eye — growths on the eyeball itself and growths on the eyelids or third eyelid. Cancer of the eyelids is often more serious; it can become malignant more quickly and can spread into the eye socket and spread rapidly, killing the cow.

Symptoms/effects: A growth on the eyeball usually appears as small white plaque near the edge of the eyeball, where the pigmented iris meets the white. Cancer on the eyelid appears along the edge of the lower lid, becoming a wart-like tumor that is pink or red, irregular in shape, bleeding easily if bumped. If the growth is on the third eyelid, it may protrude in the corner of the eye.

Prevention/treatment: Check eyes regularly and closely. If detected early, when growths are small, many of these cancers can be successfully removed by your veterinarian. Small white plaques on eyeballs can be removed if they start to enlarge (sometimes these remain small and go away by themselves). Eyelid growths are more risky; they may recur if removed. A cow with an eyelid tumor should be sold or butchered before the growth becomes malignant.

Hardware Disease

Cattle sometimes gobble up foreign objects with feed, such as small pieces of wire in the hay. This can cause serious injury if a sharp object punctures the gut lining. Bacteria from the digestive tract can then invade the abdominal cavity, creating infection that will kill the cow. Occasionally a sharp object may penetrate the heart or lungs. Prime causes of hardware disease include the following: grazing pastures with junk piles or old wire fences, eating in mangers where roofing tacks may fall into the feed, and eating pieces of wire that were shredded and chopped up by haying equipment.

Symptoms/effects: A cow with hardware disease may be dull and uncomfortable and have poor appetite. If the sharp object has already penetrated the gut wall, she quits eating and stands humped up in pain. She may have a fever or a very fast respiration rate, or breathe with grunting sounds. Her condition may quickly deteriorate; her temperature may drop as she goes into shock and she soon dies.

Prevention/treatment: Make sure there is no junk in the feed that cattle eat. Clean up any broken wires, nails, and other debris in your pasture and barnyard. Don't feed hay that might have pieces of wire or other objects baled up in it. Sometimes hay is cut near a trash dump, or broken fence wires end up in the hayfield and are baled by accident.

Since you can't always tell what's in hay you buy, one solution is to put a rumen magnet in every cow. A long, cylindrical magnet is often given to dairy cows by putting it down the throat with a non-metallic balling gun. The magnet settles in the reticulum, where it attracts and holds metal objects in that stomach, keeping them away from the stomach walls.

If a cow is already suffering from hardware disease, you may be able to save her if the problem is not too far advanced, by giving her a magnet (if she does not already have one) and antibiotics until she recovers. Your veterinarian can advise you on a course of antibiotics and treatment.

Bloat

Bloat is caused by eating highly fermentable feeds. Too much gas builds up in the rumen. Burping may not dispel the gas if it is being created faster than the animal can burp it out. Sometimes the gas is full of bubbles that won't burp up very well, or the liquid portion of the rumen contents covers the opening of the esophagus and gas cannot be burped up. Soon the tight rumen puts so much pressure on the lungs that the animal cannot breathe. Bloat is more apt to occur in animals more than a month old (with a functional rumen) than in very young calves.

A cow with bloat looks swollen on its left side, where the rumen is located.

Symptoms/effects: When viewed from behind, the animal looks full and round — "puffed up" on its left side, where the rumen is located. The animal may stand with front legs uphill, to burp easier (putting the esophagus higher so gas can escape). As the bloat gets worse, both sides become distended and the animal may stand with head and neck extended and mouth open, trying to breathe. Eventually he cannot breathe and begins to stagger and then collapses. Once the animal collapses from lack of oxygen, he dies within a few minutes.

Prevention/treatment: Avoid overfeeding types of feed that can lead to bloat, including rich alfalfa hay, excessive amounts of grain, or finely chopped hay and grain. Increase the grain in a ration slowly and do not make sudden changes in the ration. Avoid grazing lush alfalfa pastures, especially when wet from rain or dew. If you use alfalfa pastures, bloat can often be prevented by feeding a product containing an antifoaming agent (such as Bloat Guard). It can be fed in block form and given to cattle in place of a salt block. Since it tastes salty, cattle generally eat it. Start them on Bloat Guard a few days before putting them into the alfalfa field and make sure they are all eating it. Continue supplying it while they are in that field. This will reduce risks for bloat unless some animals do not consume enough of the blocks.

If cattle begin to bloat on a pasture and you herd them out, or bring one in for treatment, do not hurry them. Excessive jostling (trotting or galloping) shakes up the rumen contents and may cause or intensify bloat. Bloat can occur up to four hours after eating fermentable feeds. Even if you remove cattle from a risky pasture, check them again to be sure none are starting to bloat.

If an animal bloats badly, rumen distention must be quickly relieved. Your veterinarian can pass a tube into the animal's stomach to let out the gas. Frothy bloat does not come out easily and the veterinarian may pour mineral oil or other medications in through the tube, to break up the foam.

If bloat is so severe there isn't time to treat the animal, such as when a cow or calf collapses and is about to die, puncture the rumen to let out the gas. Use a sharp instrument called a trocar (which has a cannula to leave in the hole and keep it open so all the gas can escape), but in an emergency you can use a sharp pocketknife. Jab it forcefully into the topmost part of the distended rumen (it feels like a tight balloon on the left side of the cow), midway between the last rib and the hipbone, several inches from the backbone. The cow's skin is thick and tough, so your knife must be sharp, and you must stab very hard.

Many cows have been saved by this action. Even if you must make a large hole, your veterinarian can sew it up later. The important thing is to let gas out so the animal won't suffocate. Hold the hole open as gas rushes out, until the rumen gets back down to normal size. Then you can call your veterinarian to come stitch the animal.

Problems with Feeding Grain

Some feeding problems won't occur if cattle are on pasture or hay and never fed much grain. If you are fattening cattle on grain for butchering, however, be aware of these potential problems.

Acidosis

A large increase in grain ration can result in too much acid in the calf's body. Acidosis occurs when overfeeding on grain causes overproduction of lactic acid in the rumen.

Symptoms/effects: If not treated promptly, acidosis causes the rumen to quit working. Digestion shuts down. Then the animal may develop a fever, diarrhea, or founder (see below) and may even die. The manure becomes gray, watery, and bubbly.

Prevention/treatment: Acidosis occurs most frequently when the grain ration is increased too suddenly, but can also happen if something interferes with the regular feeding schedule. If you feed a large amount of grain, design a schedule (splitting the ration into several portions to be fed at a specific time of day) and stick to it. Problems can occur if you skip a feeding and the calf gets too hungry and then overeats at the next feeding.

Increasing the grain ration for a calf or a group of steers must be done gradually, over a period of 14 to 21 days. If you switch from a forage ration to a grain ration, make the switch gradually, taking a couple of weeks to fully increase the grain.

Acidosis may also occur if a calf goes off feed for any reason and then loads up on grain after the problem is resolved. Even something simple like manure in his water tank can make him quit drinking, and he doesn't eat as much as usual until the water is clean again.

If an animal quits eating, give sodium bicarbonate (baking soda) to neutralize the acid in the rumen. Ask your veterinarian what dosage is appropriate for the size of your animal. Mix the soda with water and give it by stomach tube or have your veterinarian do it if you do not have experience using a stomach tube.

Founder

Cattle may founder if fed too much grain, or if the ration is changed too abruptly. Founder is inflammation of the tissue between the hoof wall and inner structures of the foot.

Symptoms/effects: The feet become very sore and the animal is severely lame. The hoof wall may separate from the underlying structures, causing malformed hoofs and chronic lameness.

Prevention/treatment: The main cause of founder is acidosis, so founder usually can be prevented if cattle are not overfed on grain. Founder is always a serious emergency. Call your veterinarian.

Bladder Stones

Cattle may develop urinary stones from eating feed containing a lot of phosphorus (such as a large proportion of grain), ingesting silicates or oxalates in certain plants, or insufficient water intake. If the animal is short on fluid, salts in the urine may form crystals because the urine is so thick and concentrated. Under certain conditions, these crystals may clump together and create stones. The stones are hard masses of mineral salts and tissue cells. They can block the urinary passages and cause pain. Steers and bulls are most at risk; cows and heifers have a larger-diameter urinary tract that is less easily blocked.

Symptoms/effects: The affected animal has abdominal pain. He kicks at his belly or stands stretched, trying to urinate without success, or dribbles small amounts frequently. He may lick his belly, tread with his hind feet, swish his tail, or grind his teeth. If the stone creates a total blockage, the bladder or urinary passage may rupture. If this happens, urine goes into the abdomen, causing toxemia and death within 48 hours.

Prevention/treatment: Make sure cattle always have plenty of water, especially in winter when they drink less because of cold weather. Do not overfeed grain. Treatment for urinary stones is not always successful, but the animal may have a chance for recovery if you call your veterinarian before the animal suffers a rupture.

Skin Problems

Skin problems in cattle are often more unsightly than serious. Skin diseases caused by fungi or viruses are mainly a problem in young animals; older cattle develop some immunity.

Ringworm

Ringworm is a fungal infection, most likely appearing in winter. It is contagious, spread from animal to animal by direct contact or from halters and other equipment used on more than one animal, or by animals rubbing on the same posts or fences.

Symptoms/effects: When fungal spores are rubbed into the skin, the microscopic organism becomes established there. Hair falls out in one- to two-inch-wide circles; the skin becomes crusty or scaly. Gray scaly patches often appear on the head and around the eyes.

Prevention/treatment: It's hard to prevent ringworm if fungal spores are already in the environment. Young cattle (especially yearlings) may develop ringworm when conditions become favorable for the fungus. If you've never had ringworm on your farm, prevent it by keeping out animals that have it.

Ringworm is treated by washing the animal with a fungicide solution. If there are only a few scaly areas, you can treat them individually. Your veterinarian can recommend a good fungicide. Some types of ringworm are contagious to humans. Wear rubber gloves and wash your hands with soap immediately after treating an animal. Wearing gloves is always a good idea, since most fungicides are very strong chemicals. Ringworm will eventually go away without treatment.

Warts

Warts are skin growths caused by a virus. They often appear where skin has been broken — in an ear after an ear tag is installed, for instance. They usually appear in calves and yearlings more than adults. Mature cattle usually have developed resistance to the virus. Warts are unsightly but most will go away after a few months. Occasionally a calf will have many warts and they may become quite large and obstruct the nostrils or be easily injured. Your veterinarian can speed up the healing process with heat, freezing, or removal of the warts. If a farm has problems every year, wart vaccines may reduce the number and severity of cases.

Parasites

Cattle are affected by several types of internal and external parasites. It's easier to see external ones like flies, ticks, and lice, but the internal ones can be silent thieves, hindering weight gain by robbing cattle of the nutrition they need.

Internal Parasites

Internal parasites include worms, grubs, and liver flukes. **Worms** commonly infect cattle, especially young cattle that have not yet developed resistance and cattle grazing wet pastures. Some types of worm larvae hatch from eggs passed in manure. The larvae crawl onto forage plants to be eaten by cattle. Animals on dry rangeland are less likely to be heavily infested.

Liver flukes can be a problem in wet areas with populations of snails. Cattle eat tiny fluke larvae attached to plants growing in or near water. Young flukes migrate to bile ducts in the liver. Eggs are passed in manure and hatch in water to infest snails, where they grow into larvae to emerge and attach to vegetation.

Cattle grubs are the immature stage of heel flies. Adult flies lay eggs on the legs of cattle. The eggs hatch and burrow through the skin, then migrate through the body to lodge in the back, where they spend several months growing. They can be felt as bumps under the skin. Then they come out through the skin, making holes that damage the hide, to pupate in the ground and become flies.

Symptoms/effects: Cattle that harbor worms don't gain weight as fast as other cattle. A young animal may lose weight or have a rough hair coat, poor appetite, diarrhea, or a cough. Cattle with liver flukes may not show signs unless infestation is severe, but the liver damage leaves them more at risk for redwater disease. The bacteria that causes redwater can gain entrance via the damaged liver.

If the liver is impaired, cattle are also likely to suffer other health problems such as poisoning or photosensitization from eating certain plants since the liver is the organ that filters toxins from the bloodstream.

Animals affected by heel flies and grubs may run frantically or stand in water when adult flies are trying to lay eggs on their legs. If an animal harbors a large number of grubs over winter, they rob nutrients and reduce weight gain.

Prevention/treatment: Cattle raised in a clean place and grazing uncontaminated pastures have less worms. If they graze wet or irrigated pastures, they are at risk for heavy infestations and may need to be dewormed once or twice a year. Your veterinarian can advise you on a good deworming program (best time of year to control them) and proper drugs to use. If liver flukes are a problem in your area, use a product that kills them as well.

Cattle grubs can be controlled by killing the immature larvae before they travel to the animal's back. Treatment should be given after heel fly season is over (so there is no chance of more eggs being laid) but at least one month before grubs arrive under the skin of the back. Your veterinarian can recommend the best time of year for treatment in your region and an appropriate product to use. You can use a product that will kill lice and worms at the same time.

External Parasites

External parasites include several types of blood-sucking flies, ticks, lice, and mites. Heavy infestations reduce weight gain, spread disease, and cause great annoyance and irritation to the animal.

Lice infest cattle all year but reach peak numbers in winter during colder weather when the animal has a long hair coat. There are two kinds of lice — biting and sucking lice — and they multiply most rapidly on thin or sick animals.

Symptoms/effects: Lice cause itching and rubbing. The animal may rub out patches of hair, especially over neck and shoulders. If you look closely at the skin you can see the lice, especially around the eyes or muzzle. Heavy infestations of lice cause cattle to lose weight, lower their resistance to disease, and may result in death from anemia.

Prevention/treatment: Lice are spread from animal to animal. Do not introduce new animals into your herd without delousing them first. If your cattle have lice (and most cattle do), treat them in the fall before lice populations build to high levels. Some products remain effective long enough to kill the remaining eggs that hatch after treatment. Other products only kill adult lice and you need to repeat the treatment two weeks later to kill the young ones.

Consult your veterinarian regarding treatments. Follow label directions on all insecticides. For best lice control, treat cattle in the fall and again in mid-winter or early spring.

Mites are tiny parasites that feed on skin and deposit their eggs there. Scabies mites deposit eggs on the skin and attack any part of the body covered with hair. Mange mites feed where hair is thin and skin is tender, such as the inner surface of the legs or the top of the udder. They burrow into the skin to lay their eggs.

Symptoms/effects: Scabies mites prick the skin to obtain food from tissue fluids. This causes intense itching, inflammation and sores. Oozing from the sores mixes with dirt, creating a gray or yellow scab that becomes infected and hardens. The scabs increase in size as mites feed on withers, shoulders, or tail head, eventually affecting the whole body. Itching is so intense that cattle are always rubbing instead of eating.

Mange mites also cause itching. If not controlled, they spread under the belly to the brisket and upward toward the tail, eventually covering the body and causing hair loss and heavy, dry scabs. The skin has a thick, wrinkled look.

Prevention/treatment: Both types of mites are spread by direct contact between animals or contact with objects that have been rubbed by infested cattle. Do not bring infested cattle onto your place. To eliminate mites, thoroughly wet skin with the proper insecticide (two applications, 10 to 12 days apart). Dewormers containing ivermectin, moxidectin, or doramectin kill most external parasites, including mites.

Ticks are a problem in many regions. They spread diseases as well as cause anemia and weight loss in cattle. There are two types: hard-bodied ticks like the wood tick (also called a dog tick) and the deer tick, and the smaller, soft-bodied ticks. Tiny deer ticks spread Lyme disease.

Symptoms/effects: Heavy infestations of ticks cause severe irritation, weight loss, and sometimes anemia and death. Hard-bodied ticks feeding at the base of the skull can cause tick paralysis in cattle.

Prevention/treatment: Ticks can be controlled with certain insecticides used as sprays, dips, or powders. Ear ticks can be controlled by use of insecticide ear tags or by applying the proper insecticide into the animal's ears. Consult your veterinarian.

Flies bite and suck blood, spread disease, and cause great annoyance. The reduction in grazing time (due to the annoyance and irritation) and blood loss can reduce weight gains.

Symptoms/effects: If cattle spend all their time and energy trying to get away from flies or swishing them off, they don't graze enough and lose weight. Horn flies may reduce weaning weights of calves by as much as 40 pounds.

Prevention/treatment: Most flies can be controlled with the proper insecticides, which can be applied several ways. Periodic application of sprays and pour-on insecticides will reduce flies. Dust bags or back rubbers can be put in gateways or areas where cattle will use them every day; when an animal rubs on this device, it applies the insecticide. Horn flies and face flies can be well controlled with insecticide ear tags, put into the ears in early summer. The tag continuously releases insecticide as it is rubbed against the hair; the animal rubs it over his body as he reaches around to lick or rub himself.

Giving Injections

Many medications and most vaccines are given by injection with a syringe and needle. Many injections are given intramuscularly, deep into a big muscle. Others are given subcutaneously, between skin and muscle. A few are given intravenously, directly into a large vein. Veterinarians should give intravenous (IV) medications, but you can

PROPER RESTRAINT

Always restrain an animal before giving an injection. Put a large animal in a chute. A small calf can be pushed into a fence corner and held securely against the fence. If it is merely tied to a fence, it may still move around too much or kick you. Don't stand behind or beside the animal unless it is properly restrained — so it cannot move around or kick — with a panel between you and him.

learn to give intramuscular and subcutaneous injections. Have an experienced person show you how to fill a syringe, measure dosage, and give the injection.

Intramuscular Injections

Intramuscular (IM) shots should be given in the thickest muscle of the neck, if possible, to avoid damage and scarring in the best cuts of meat (rump and buttocks). Sometimes an injection causes a local reaction and a knot in the muscle, or even a small abscess. It's better to have this occur in the neck, where it's more easily trimmed out during butchering.

Make sure the area where you will put the needle is very clean, without mud or manure, or bacteria might get into the muscle and create an abscess. Wet skin and hair increases the risk.

Detach the needle from the syringe, unless it is a "gun"-type syringe. Press the area firmly with your finger or the edge of your hand before putting in the needle. This desensitizes the skin and the animal will not be so startled (and jump!) when you press in the needle. Put the needle in with a forceful thrust, so it will go through the skin and into the muscle. A new, sharp needle goes in with least effort — and less pain to the animal — than a dull one.

The advantage of putting the needle in by itself, before you give the shot, is that if the animal jumps, you can wait until he settles down and relaxes before you give the injection. Then you won't be squirting medication or vaccine in as the animal jumps, risking loss of part of the dose. You can also tell if you've hit a vein before giving the injection. Do not put an intramuscular shot into a vein. If the needle starts to ooze blood before you attach the syringe, take it out and try a slightly different spot.

Administering an IM injection in the neck.

Subcutaneous Injections

Subcutaneous (SQ) shots are given under the skin. The easiest way is to lift up a fold of skin on the shoulder or neck, where the skin is slightly loose, and slip the needle in. Aim it alongside the calf so it goes under the skin you have pulled up, and not into the muscle. This is easy to do with a "gun" type of syringe, without taking the needle off. If using a disposable syringe, you can put the needle in first, then attach the syringe.

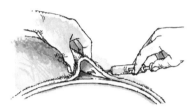

Administering an SQ injection.
(cutaway view)

After giving an injection, discard the syringe and needle (if they are disposable) in a safe container in the trash can to avoid accidental needle pricks. If they are reusable, boil them before the next use. Syringes can also be taken apart and boiled.

Oral Medications

When giving pills or liquid medication by mouth, restrain the animal in a chute with a head catcher, or tie its head to the side of the chute so it cannot swing its head away or hit you with its head while trying to avoid the medication.

Pills

Pills and boluses can be given with a balling gun. A balling gun is a long-handled tool that holds the bolus while you put it toward the back of the mouth. When you press the plunger it pushes the pill out of its slot into the animal's throat to be swallowed. If you aim it far back, the animal must swallow the pill when the tool releases it. The tool keeps your fingers from being crushed by the animal's teeth.

Be careful when giving pills or examining the inside of an animal's mouth. Cattle have no top teeth in front but can crush your fingers between the molars if your fingers are back too far when the animal bites down.

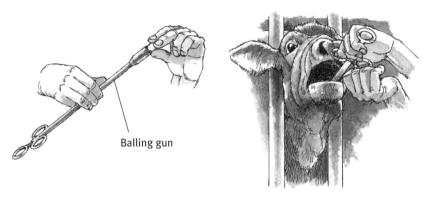

Balling gun

A balling gun enables you to place the pill at the back of the throat, so the animal has to swallow it when you press the plunger.

Liquids

Giving liquid medications (or pills dissolved in water) to a calf is easy with a large syringe (minus the needle) or a special dose syringe with a metal tube that goes to the back of the mouth. Fill it to the proper dosage and slowly squirt the medication into the back of the mouth. Position the syringe into the corner of the mouth and aim it far back so the calf must swallow the medication.

If you're giving a large dose (such as several ounces of Pepto-Bismol or Kaopectate), do not squirt it all at once; he may be able to spit some back out, or he may choke on it. Squirt a little at a time, allowing him to swallow each portion before you squirt in more. Keep his head tipped up so medication can't run back out of his mouth. You can refill the syringe as needed for a large dose.

A small calf can be backed into a fence corner and his head held securely between your legs as you do this, whereas a large calf must be restrained more fully in a chute or with someone helping you hold him. It's difficult to give liquid medication to an adult animal unless you use a stomach tube, and you will need your veterinarian's help to be sure it goes down the throat and not the wind pipe.

Proper Care Prevents Illness

By giving good care and paying close attention to cattle, you can make sure they stay healthy. You can detect any problems such as injuries or signs of illness early on and can deal with those before they become serious.

Keep cattle comfortable and avoid stress. To prevent many types of illness, try to avoid stress and feed cattle properly. Make sure they have shelter from cold and windy weather and shade during hot weather. Stress from weather extremes or from inadequate feed or water may lower the animals' resistance to disease.

Cattle suffer as much stress in hot weather as in cold weather. A heat wave, with high humidity, can cause heat stroke and death in cattle. During very hot weather you might have to install a fan in a barn stall, for calves or dairy cows, or hose cattle down with a misty wet spray from your garden hose if they are confined in a pen without access to shade.

Sanitation is important. A clean environment is always beneficial. It is easier to try to prevent infectious diseases than to treat them.

If there are several calves in a pen or barn, or if there have been cattle there before, there may be bacteria and viruses lurking. You can help prevent diseases and infections by thoroughly cleaning and disinfecting facilities between calves (especially when raising young dairy calves) or between groups of calves. Get rid of all old bedding and scrub the walls and the floor, if there is one, of each shelter with a good disinfectant. Your veterinarian, extension agent, or a dairyman can recommend an appropriate product. Even in a large pen or pasture, cleanliness is important. Cattle need clean areas to eat and to lie down, even if it means putting out straw for bedding and using feed racks during muddy times of the year.

CHAPTER FOUR

Choosing a Beef Breed

THERE ARE MANY BREEDS OF CATTLE TODAY. They have different characteristics in size, color, muscling, milking ability, and weather tolerance. Some are long-haired and better for cold climates; some are short-haired and have more sweat glands, which is best in hot climates. Some have horns and some don't.

Christopher Columbus brought cattle to the New World in the late 1400s. The Vikings may have brought some earlier, but those were probably butchered for food. Spanish explorers and settlers brought the ancestors of Longhorn cattle. Later the Pilgrims and other settlers along the Atlantic Coast brought cattle from England.

Beef breeds in North America are descendents of cattle imported from the British Isles (England, Scotland, and Ireland), Europe ("continental" breeds), and India. Many modern breeds are mixes of these imported breeds. Those first cattle from Spain were soon outnumbered by British breeds, as the American colonies expanded westward. Herefords from England were also used to improve beef production from western longhorn cattle. Later, during the 20th century, larger "continental" cattle were imported to again increase the size of some American beef animals.

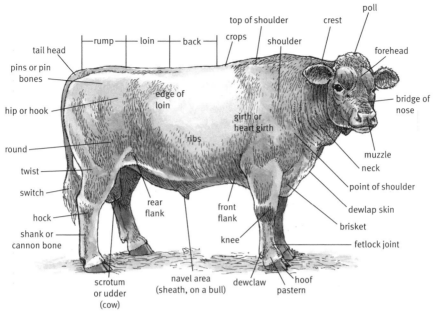

Labels on the diagram:
poll, crest, top of shoulder, shoulder, forehead, rump, loin, back, crops, tail head, bridge of nose, pins or pin bones, edge of loin, hip or hook, girth or heart girth, muzzle, round, ribs, neck, twist, point of shoulder, switch, rear flank, front flank, dewlap skin, hock, brisket, shank or cannon bone, knee, fetlock joint, scrotum or udder (cow), navel area (sheath, on a bull), dewclaw, hoof pastern

PARTS OF A BEEF ANIMAL

What Is a Breed?

For hundreds of years, cattle have been selected and bred for certain purposes. They were originally wild animals. There were only two species — the Aurochs of Europe *(Bos taurus)* and the Zebu of Asia, India, and Africa *(Bos indicus),* the hump-backed, droopy-eared cattle of the tropics. Most breeds in North America are *Bos taurus.* The Brahman is the most well-known Zebu breed. See pages 235–250 for a photo gallery of popular beef and dairy breeds.

How Breeds Began

After cattle were domesticated, stockmen noticed some were better than others in certain traits. For instance, if a cow gave more milk, she raised bigger calves. They learned they could improve on certain traits if they mated animals with desired characteristics. If they chose a bull whose mother produced lots of milk and bred him to cows that gave lots of milk, the daughters from those matings were better milk cows. Other cattle were bred to be big, strong oxen for pulling carts. These were ancestors of beef breeds with lots of muscle and larger carcasses.

Breed Registries

During the past few centuries specific breeds were developed — the Hereford in England, Galloway and Aberdeen-Angus in Scotland, Charolais in France, Simmental in Switzerland, Chianina in Italy, and so on. Each breed has characteristics that distinguish it from other breeds.

The stockmen who raised a certain type of cattle got together to create a registry for animals in their breed, to keep track of bloodlines and pedigrees to maintain the purity of the breed. They formed organizations to write up standards for members to follow when breeding or selecting animals. Over the years new breeds have been created by combining some of the older breeds and selecting for more specific traits. Today there are more than one billion cattle in the world. There are more than 100 different breeds in the United States. A few are dairy breeds, but most are beef cattle.

British Breeds

The breeds we call British are the ones that originated in the British Isles — England, Scotland, and Ireland.

HEREFORD (photo p. 243)

The Hereford breed began when white-faced red cattle of Holland were mated with smaller black cattle native to England. Farmers of Herefordshire were selectively breeding the offspring of this cross by the mid-1700s. They selected for the red, white-faced coloration; there were no more black animals in that breed.

Herefords were brought to America in 1817 and soon became the most popular breed in the early livestock industry. Hereford bulls were bred to Texas Longhorn cows (descendents of early Spanish stock brought to the New World in the 1500s) to upgrade western range cattle. Herefords were also crossed with farmers' Durham (Shorthorn) milk cows to produce beefier calves.

The Hereford is well known for its red body and white face. The feet, belly, flanks, crest (top of neck), and tail switch are white. They also have a large frame, big bones, and a mellow disposition.

POLLED HEREFORD *(photo p. 245)*

The Polled Hereford became a separate breed in the early 1900s, after a few stockmen began breeding the offspring from hornless mutations. They created their own registry and breeding goals. Now the Polled Hereford is a distinct breed with only subtle differences beyond the difference of hornlessness.

ANGUS *(photo p. 238)*

Aberdeen-Angus originated in Scotland when farmers crossed native cattle of two counties (Aberdeenshire and Angus), resulting in black cattle that were genetically polled (hornless). Angus is one of the few breeds originating only for beef purposes and not for milk or draft; they roamed hill country and were wilder and smaller than Herefords.

In early years they were known for calving ease (small calves at birth), which made them popular for crossing with larger, heavy-muscled cattle. They are noted for early maturity, marbling of meat (flecks of fat to make it tender and juicy), and motherliness. Angus cows protect their calves aggressively and are hot-tempered. Brought to America in 1873, Angus are now very popular because of their fast finishing in the feedlot, lack of horns, and maternal qualities (high fertility and lots of milk for their calves).

RED ANGUS *(photo p. 245)*

Red Angus became a separate breed in the 1950s. Red color crops out in black herds when calves inherit a recessive red gene from each parent. Founders of the Red Angus breed collected some red cattle and started a new breed. They had goals for high levels of production and breed standards; Red Angus now have a uniformity and differences besides color that set them apart from black Angus.

SHORTHORN *(photo p. 238)*

Shorthorns were developed by crossing the early cattle of England with cattle from northern Europe, creating dairy animals originally called Durhams. Scottish breeders later selected for a more compact, beefy type — the ancestors of beef Shorthorns in North America today.

There are also milking Shorthorns, now a separate breed. The Shorthorn was one of the earliest breeds in the U.S., arriving in the 1780s.

The cows have good udders and give a lot of milk. Even though calves are born small (few calving problems), they grow big quickly. Shorthorns can be red, roan, white, or red-and-white spotted.

GALLOWAY *(photo p. 242)*

Galloway is one of the oldest breeds. Vikings brought its ancestors to southwestern Scotland in the ninth century. Scottish Galloways first came to North America in 1866. They are polled and hardy, with a heavy hair coat in winter. Cows live a long time, often producing calves until age 15 to 20. Calves are small and easily born, but grow fast. Most Galloways are black, but some are red, brown, and white with black ears, muzzle, feet, and teats, or belted (black with white midsection).

SCOTCH HIGHLAND

Scotch Highland cattle are small, with long shaggy hair and big horns. They lived in rugged Scottish highlands for hundreds of years and were first brought to North America in 1882. These hardy cattle do well in cold weather. Shaggy coats also give protection from insects; their long forelocks protect their eyes from flies.

DEXTER *(photo p. 241)*

Dexter cattle were developed in Ireland in the 1800s and brought to North America in 1905. They are probably the smallest cattle in the world used for milk and beef. The average cow weighs less than 750 pounds and is only 36 to 42 inches tall at the shoulder. Bulls weigh less than 1,000 pounds and are 38 to 44 inches tall. They are quiet, easy to handle, and the cows give very rich milk.

Continental Breeds

Many European breeds in America, imported in the past 40 years, add size and muscle (and sometimes milk) to cattle in North America. Today we have our own versions of these breeds.

CHAROLAIS *(photo p. 240)*

Charolais (imported in 1934) are white, thick-muscled cattle, originating in France as draft animals. They were the first continental breed popular in America for crossbreeding, since they were larger than British breeds.

SIMMENTAL *(photo p. 246)*

Simmental is one of the oldest and most widely used breeds around the world. Originating in Switzerland, these yellow-brown cattle with white markings have rapid growth and good milk production. They came to North America in 1967.

LIMOUSIN *(photo p. 243)*

Limousin cattle are red and well-muscled. Developed in western France, they were first imported in 1969, becoming very popular in America. Cattlemen like this breed's abundance of lean muscle and the fast growth of calves.

TARENTAISE *(photo p. 247)*

Tarentaise are cherry red, with dark ears, nose, and feet. They originated in the French Alps and are related to Brown Swiss. First bred for milking, they became moderate-sized dual-purpose cattle used for milk and meat. They mature early, are comparable in size to British breeds, and were first brought to North America in 1972.

SALERS *(photo p. 246)*

One of the oldest European breeds, Salers cattle are from south-central France. This breed is horned, dark red in color, and popular in America for crossbreeding because of good milking ability, fertility, calving ease, and hardiness.

CHIANINA

Chianina are large white cattle — the largest in the world — and the oldest breed in Italy (since before the Roman Empire). Developed as a draft animal, Chianina can be six feet tall at the shoulder and may weigh more than 4000 pounds. They were brought to North America in the early 1970s and used for crossbreeding.

GELBVIEH *(photo p. 242)*

Gelbvieh came from Austria and Germany and were first used for draft, meat, and milk. This breed is light tan to gold color. The calves grow fast, and heifers mature quicker than those of most other continental breeds.

OTHER CONTINENTAL BREEDS

There are many other European breeds in North America today, including **Maine Anjou, Normande, Pinzgauer, Piedmontese, Braunvieh,** and **Romagnola.** Continental cattle are usually larger, leaner, and slower to mature than British breeds.

Breeds from Other Places

Some of our American breeds came from places other than the British Isles and the European continent.

MURRAY GREY

Murray Grey are silver-gray cattle from Australia, dating back to a single Shorthorn cow that produced 12 gray calves between 1905 and 1917 when bred to Angus bulls. This breed is becoming popular because of its gentle disposition, moderate size, and fast-growing calves. Calves are small at birth but reach 700 pounds by weaning.

BRAHMAN *(photo p. 240)*

Brahman cattle originated in India and are easily recognized by the large hump over neck and shoulders, loose floppy skin on the dewlap and under the belly, large droopy ears, and horns that curve up and back. They come in a variety of colors.

American Brahman cattle were developed in the Southwest from several strains of Indian cattle imported between 1854 and 1926, and some from Brazil. Brahmans do well in the South: They tolerate heat and humidity and are resistant to ticks and other hot climate insect parasites. They are large animals. Calves are very small at birth but grow rapidly because the cows give lots of rich milk.

American Breeds

Several American breeds have been developed by crossing other breeds or selectively breeding the cattle brought over in early times.

TEXAS LONGHORNS *(photo p. 247)*

Texas Longhorns are descended from wild cattle left by Spanish settlers in the Southwest. After the Civil War, ten million were trailed north to market. By the early 1900s increasing numbers of British cattle reduced demand for Longhorns; they nearly disappeared until a few stockmen began breeding them again. They are moderate-sized, known for calving ease, hardiness, long life, and high fertility.

BRAHMAN-BASED BREEDS

Several American breeds were created by crossing Brahman with other cattle, to develop heat-tolerant breeds with good beef production. **Santa Gertrudis** cattle were developed at the King Ranch of Texas by crossing Brahmans with Shorthorns. **Brangus** are a mix of Brahman and Angus. **Beefmasters** are a blend of Brahman, Shorthorn and Hereford. There are also **Charbray** (Brahman-Charolais) and **Braford** (Brahman-Hereford).

Crossbreds

Cattlemen often cross breed cattle to create herds with the traits they want. A crossbred is an animal with parents of different breeds. Cross-breeding is a very useful tool for the beef producer.

Cattle are raised in a wide variety of situations and environments, from lush green pastures to dry deserts and steep mountains, from cold northern climates to hot and humid southern areas. Each farmer or rancher tries to utilize or create a type of cattle that will do well and raise good calves in his own situation.

No single breed is best in all traits that are important in beef production. Many people crossbreed because no one breed can offer all the advantages that might be needed. It takes a different kind of cow to do well in desert country, with many miles to travel between the sparse grasses and water, than one that will perform best on lush alfalfa pasture or corn stalks.

The most effective genetic advantage in raising cattle is taking advantage of hybrid vigor — the result of mating two animals that are very different. This is only obtained through crossbreeding. Hybrid vigor increases fertility, milk production, life span, vigor, health, and many other important traits. With careful crossbreeding (selecting good individuals from each breed, rather than poor ones), the stockman can produce cattle that outperform the parent breeds. Good crossbred females make the best beef cows.

Composite cattle are a blend of different breeds into a uniform type of crossbred. Several composites were created in the past 30 years and more new ones are being formed. Composites are not a new idea; nearly every breed we have today began as a composite, followed by many generations of selective breeding to standardize certain characteristics such as the genetic base of that breed.

A composite takes advantage of genetic traits from more than one breed, combining them into one animal. Brangus, Beefmaster and Santa Gertrudis are examples of successful composites that are now considered breeds. More recent blends are the **Hays Converter** (a Canadian breed made up of several beef breeds combined with Holstein—a dairy breed), the **RX3** (a blend of Hereford, Red Angus and Holstein), and numerous blends of various beef breeds.

How Shall I Choose?

There are so many breeds and crosses to choose from! Before selecting cattle of a certain breed or combination of breeds, you should consider several things:

Evaluate strong points and faults. All breeds have strong points and shortcomings. No one breed is perfect for all situations, and this is why crossbreds and composites are becoming more popular. There can be vast differences among individuals within a breed, so examine each animal carefully before you buy it.

Take local conditions into consideration. When selecting cattle, try to make an objective decision about breeds. Evaluate the region where you live to see which breeds are best suited for your conditions. A woolly breed might not do well in a hot climate and a Brahman or Santa Gertrudis might not do well in the cold.

Choose between registered purebreds and commercial cattle. If you want to start a breeding herd, decide whether you want registered purebreds or commercial cattle. Commercial cattle can take advantage of the benefits of crossbreeding, and are usually the best cows. If you want to raise cattle as beef to butcher or to sell to your neighbors as beef, good crossbreds may be your best bet.

If you want to raise registered purebreds, select a breed being raised in your area, so you can find bulls in a herd close to home, unless you want to depend on AI (artificial insemination; see chapter 9). You can buy heifers or cows from a local breeder. Most purebred breeders are glad to help someone else get started, and even if you just buy one or two heifers from their herd, they will probably help you breed them.

Whether you are looking for heifers, a few cows, or beef steers, there are people who can help you find them. Your county Extension agent can tell you which breeds and crosses are being raised in your area, and can help you contact the stockmen who are raising the kind you'd like to buy.

CHAPTER FIVE

Selecting and Transporting Beef Cattle

ONCE YOU HAVE PREPARED A PLACE FOR CATTLE, by checking fences on your place or creating a pen or a fenced pasture, you can purchase the animals and bring them home.

Where Can I Buy a Calf or Cows?

You can buy calves or cows at an auction sale or at a farm or ranch if the owner wants to sell some. Purchase weaned calves in the fall to raise for butchering or resale as yearlings. A local purebred breeder may have an annual sale of bred heifers if you are interested in raising purebreds, or you can buy cows at a dispersion sale. Read the ads in your local farm papers to find out where to buy cows or heifers.

One advantage to buying at a production sale or by private treaty directly from the person who raised the animal is the chance to learn about the cattle. You can ask about their medical history — what vaccinations they've had, when those were given, and how soon they will need the next vaccination. Calves will need several vaccinations against certain diseases. If they haven't already had these injections, they should have their shots soon after you bring them home.

A big disadvantage to buying cattle at a livestock auction is the lack of information, such as where the cattle came from and their health and vaccination history. There is also risk that they came into contact with illnesses at the sale yard and might be sick after you bring them home.

The Cattle Market

Calves, and any animals going into feedlots or for slaughter, are sold by the pound. Brood cows and bred heifers are generally sold by the head. If you want to buy weaned calves, before you go to an auction or visit a farm or ranch to look at cattle, find out the per-pound price of beef animals. If you want breeding stock, find out what bred heifers or cows are bringing by the head. Prices change a little every week, so stay up to date on the current market. You don't want to pay more than market price, especially if you are hoping to make a profit in your cattle-raising venture.

Understanding Cattle Prices

Steers bring several cents more per pound than heifers of the same weight because their carcasses are more valuable as meat. For instance, if a 550-pound steer brings 90 cents per pound, a similar heifer would be worth about 82 to 85 cents. Yearlings bring less per pound than calves. A 1000-pound yearling steer might be worth 75 cents per pound when a 550-pound steer calf is worth 90 cents. The yearling brings more total dollars because he weighs so much more. The 1000-pound yearling at 75 cents per pound is worth $750 while the 550-pound calf at 90 cents is worth $495. If you pay $450 for the weaned steer, you can still make money during the year you keep him and feed him, if he gains well and feed costs do not exceed the price difference between what you paid for him and his later selling price.

Breeding stock prices are based on the animal's potential offspring over the course of her reproductive life. The price for bred heifers is often higher than the price for older, bred cows. The heifer probably has a longer life of production ahead of her. Young cows (age three to four) just hitting their peak of production cost more than older cows.

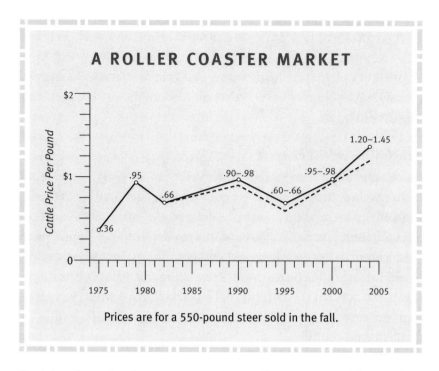

A ROLLER COASTER MARKET

Cattle Price Per Pound

.36
.95
.66
.90–.98
.60–.66
.95–.98
1.20–1.45

1975   1980   1985   1990   1995   2000   2005

Prices are for a 550-pound steer sold in the fall.

Registered purebred cows may cost more than commercial cows. In a year when a bred heifer is worth $800, young bred cows might be worth $600 to $700, while older pregnant cows may be worth only $500. Prices will vary from year to year.

Often the most expensive category is pairs (cows with young calves at their side). Cow-calf pairs may cost a little more than bred heifers. Yet this may be the most economical group to purchase, especially if the cow is already bred again and pregnant with another calf; you are getting a three-in-one package. The bred heifer is also an unknown quantity, and more of a gamble. You don't know whether she will calve easily or need help, nor whether she will be a good mother or breed back again.

One way to start inexpensively, if you have a lot of pasture and don't need to buy much hay, is to purchase older cows cheaply — pregnant cows, or cows with calves at their side — and keep the heifer calves. Even if the older cows only stay on your place long enough to have one or two more calves, some of their calves will likely be heifers and you can build your herd with those.

Selecting Beef Calves

When buying calves to raise for beef, most people choose steers, even though heifers are cheaper. Steers gain faster and grow larger, and are not as flighty or ambitious. They are less apt to crawl through fences. Beef calves raised for consumption or sale as butcher beef, or for resale as yearlings, needn't be purebred. The best beef steers are usually crossbred, since crossbreds tend to grow faster, with better feed conversion (more pounds of beef produced on less feed) than most purebreds.

A purebred is an animal of a certain breed that has no other breeds in its ancestry. The term should not be confused with Thoroughbred, which is a certain breed of horse. A purebred is not necessarily registered. The animal may be a purebred that was not registered. A registered purebred has a registration number, recorded in the herd book of a breed association. The association gives the owner a certificate stating the animal is the offspring of certain registered parents.

Type

Select a type to fit your purpose. The best kind of steer to raise for beef is usually a fast-growing, well-muscled animal that will reach market weight of 1050 to 1300 pounds by the time he is 14 to 20 months of age. If you plan to butcher the animal for the family freezer, it might not matter if he is a little smaller or larger at his finishing weight, but if you want top price when you resell him, he should be of moderate frame size — the ideal type to go right into a finishing lot for a short time on grain. Finish weight is the term for the weight at which a beef animal has enough body fat and is ready to be butchered.

If you are raising a group of steers, it pays to select some that are fairly uniform in size and flesh condition, so they will be ready to sell at the same time. Otherwise, a buyer might cut some back or give you a lower price per pound for some of them. If you plan to sell the steers individually to friends for beef, this won't matter; you can sell each one when he is finished, so that he will have enough body fat to be good eating.

Frame Score

Beef cattle are categorized by "frame score," which tells whether they are small-, medium-, or large-bodied. A small-framed animal will mature at a lighter weight and will not produce as much meat on his small carcass as a larger-framed animal. If you try to push him to a bigger size, he will merely become too fat. He is not genetically capable of growing larger.

A large-framed steer, by contrast, may grow too big before he gets fat enough to butcher, taking longer to mature, and hence requiring too much feed. The most profitable and practical kind of beef steer, in most situations, is one with a medium frame, and these bring the best price per pound if you are selling them.

Conformation

The calves you pick should have a lot of muscle, not just fat. A calf should have nice, smooth lines, and should not be sway-backed or pot-bellied. He should have a deep body, not shallow. He should be long and tall, but not extremely tall. A short-legged calf may quit growing before reaching ideal market size and weight. A really long-legged calf may keep growing too long (due to large frame size) or be too leggy, without enough body mass and muscling.

Select a calf with good conformation and structure. You want him to have a good shape so he can "put meat on it." A calf that is too short in the back or hip will not produce as much meat. He also needs strong feet and legs so he can travel well and won't become injured or crippled.

When selecting beef calves to buy, make sure they look healthy and have moderate flesh covering. Thin calves may be sick or might not have been fed adequately and may be stunted due to deficiencies; it's better not to gamble on them. Calves that are very fat may not grow well; they might finish too quickly. It's usually most profitable to buy calves in moderate body condition. They won't be as expensive as heavy calves and will usually grow fast. Even if you have to pay more per pound for the lighter ones, they'll still cost less total dollars because they aren't as big.

FIGURING THE FRAME SCORE FOR BEEF CALVES

To figure a calf's frame score, measure his height from the ground to the top of the hip when the animal is standing squarely. Look up his age on this chart and find the hip height on that age line. The top of the column gives the frame score. For instance, a 10-month-old calf that is 45 inches tall at the hips would be a frame score 4. You won't be restraining beef calves for measuring when you are looking at them to select those you want to buy, but someone knowledgeable about cattle can help you estimate their frame score.

AGE (months)	FRAME SCORE (hip height in inches)						
	1	2	3	4	5	6	7
5	34	36	38	40	42	44	46
6	35	37	39	41	43	45	47
7	36	38	40	42	44	46	48
8	37	39	41	43	45	47	49
9	38	40	42	44	46	48	50
10	39	41	43	45	47	49	51
11	40	42	44	46	48	50	52
12	41	43	45	47	49	51	53
13	41.5	43.5	45.5	47.5	49.5	51.5	53.5
14	42	44	46	48	50	52	54
15	42.5	44.5	46.5	48.5	50.5	52.5	54.5
16	43	45	47	49	51	53	55
17	43.5	45.5	47.5	49.5	51.5	53.5	55.5
18	44	46	48	50	52	54	56

Disposition

The general nature of cattle is also important when you make your selections. Just like humans, cattle have different personalities — nervous or calm, suspicious or trusting, friendly and curious, or independent or mean. Evaluate the attitude of every animal you consider.

Some animals are more placid and easy-going than others. They are easier to handle and manage than wild, nervous animals. Calm cattle will be less apt to run through fences when you try to move them from pasture to pasture or bring them into the corral. They also gain weight better; they spend more time eating and less energy worrying or pacing the fences.

If you are buying just one or two calves, be doubly sure to choose mellow individuals and not flighty ones that might "go bonkers" by themselves. A wild, snorty calf is a risk, even if he is big and beautiful. Wild cattle are dangerous; if they become frightened they act first and think later, and could hurt you if they kick you, knock you down, or charge into you in their efforts to get away when you come too close. Choose gentle, smart animals that will be a pleasure to have around. They are more fun, less work, and more profitable.

Getting Them Home

Make sure you have a safe place to put cattle when you first bring them home. It's always wise to put them in a strong corral for the first day or two, or up to one week if they've just been weaned, until they become accustomed to being in a strange place before you turn them out on pasture. If you don't have a truck or trailer to transport cattle, make arrangements with someone who does. If you are buying the animals from a farmer or rancher, he may be able to haul them for you. It is difficult to load cattle alone. If you bought cattle from a farmer or rancher, he will probably help you load them. When loading cattle into a trailer after an auction, people at the auction may help you herd them into the trailer from a loading alley.

A trailer often works best for hauling and loading or unloading cattle because it is low to the ground. The cattle can just step out of it. A truck or pickup with stock rack may be too high for cattle to jump out without risk of injury. If you don't have a loading chute, back up against a high bank or use a ramp of sturdy boards for cattle to walk down from the truck.

When unloading cattle, make sure the trailer is backed up into the pen far enough that they can't go anywhere except into the pen. Swing the pen gate tightly against the trailer so there is nowhere else for the animals to go but into the pen. A frightened animal may try to bolt through a very small opening. Don't stand in a spot where he might run over you.

If you bought newly weaned calves, remember they probably lived with their mothers in large pastures. Some calves haven't seen humans up close except when they were herded into the corral to be weaned or sold, or for vaccinations.

They may have been calm when you looked at them on the farm with their herdmates, but now they are upset and frightened. They may try to run over you if you are in the way as they come out of the truck or trailer. Work quietly and calmly and stay out of the way when you open the door or tailgate. Let them go into the pen and discover their new boundaries. Give them some feed and water. If they have a basically calm disposition they will soon settle down and start eating.

Safety first: When working with cattle, make a habit of trying to think like they do, to avoid problems and accidents. Cattle are large, strong animals and can easily hurt you if you get in their way. Don't expect them to be calm and gentle at first. When brought to a strange place, they may need some time to settle down and lose their fear, and come to recognize you. Once they know you — and realize you are their meal ticket, bringing them feed or moving them from pasture to pasture for fresh grass — they will become very tame and it will be a lot easier for you and for them. Take your time at first, and let them adjust to their new place and to you.

CHAPTER SIX

Care and Handling of Beef Cattle

To PROPERLY CARE FOR CATTLE, you must know something about feeding them (see chapter 2) and how to handle them. Part of being a good stockman is understanding how cattle feel and think — why and how they react to various situations. If you understand them and handle them accordingly, it becomes a lot easier to work with them safely and manage them in ways that are most beneficial to their health and well-being.

Understanding Cattle Behavior

To understand your new cattle, try to think like they do. Cattle are herd animals. This means they live together in groups. Even though they have been domesticated for thousands of years, they are still happiest when they can live in a large family group like their wild ancestors did.

Wild cattle stay in a herd as protection from predators. In prehistoric times their main enemies (besides humans who hunted them for meat) were wolves. A lone cow had no chance against a pack of wolves, even with her sharp horns. While she was busy trying to fight off the wolves in front, wolves could grab her from behind and slash the tendons in her hind legs with their teeth, so she couldn't run.

Cattle had a better chance of survival if there were many sets of horns. When wolves approached, cows would bellow and come running together, to form a tight group. Adult cattle made a circle around the young ones to protect them. All the cows faced the wolves, with their horns ready to jab at any wolf that came too close.

Calves grow up learning that the safest place to be is with the herd. Even if cattle scatter out to graze, they usually keep in sight of one another. The ancient instinct to face danger together is still strong today. If something frightens cattle, they group together. Their first reaction to strange dogs or people is the same.

Yearlings and young cattle like to wander and explore new territory, but they travel in a group. If one goes to water, they all go. When the leader of the group decides it's time to get up from a nap and go graze, they all go. It's not that they are copycats. They instinctively act like groupies for protection.

A newly weaned calf misses his mother and his herd, and may be afraid of you. Try not to startle him, or he may become so upset that he could crash into the fence. Cattle are happiest when they are with other cattle. A calf is especially insecure if he's all alone; he wants to be with adults. This is why he becomes so frantic if he is suddenly taken away from his mother (and the herd) and weaned.

After you bring cattle home, they must become familiar with you and not be afraid. This may not take a while if they are already tame and gentle. It may take several days or weeks if they are not accustomed to people being close to them. If they are timid and insecure, work very calmly, slowly, and quietly around them until they learn to trust you. Try not to startle them with loud noises or sudden movements or they may run from you.

Weaning Is Stressful

If calves were already weaned before you bought them, they won't be quite so unhappy and frantic. They've already gone through the emotional panic and stress of losing their mothers and have resigned themselves to being without them. They won't be so desperate to escape from the pen or pasture to go find their mothers. They just need time to adjust to their new home.

Two or more calves always creates a better situation than one calf. If at all possible, acquire at least two animals. A lone animal is more stressed because it feels so insecure. A newly weaned calf will be frantic by itself, but even yearlings or adult cattle are happiest with a buddy. If you buy just one calf, it is *very* important to choose one with a mellow, easy-going disposition. A wild one will be extremely emotional and hard to handle when he's in a panic. A placid, gentle calf will make a better beef animal than a calf that's flighty and still thinking of people as predators — something to run away from or fight.

Treat weaning calves with care. If they were taken from their mothers when you brought them home, it may take several days of stressful adjustment. They may pace the fence and bawl, with little interest in feed or water. This stress is hard on a calf and makes him more vulnerable to illness since stress taxes the immune system. If the weather is cold, rainy, or windy, a calf going through weaning is at risk for pneumonia.

Weaning is stressful because calves are suddenly deprived of mama's milk, but the emotional stress of being separated from mama is even worse. A weaned calf has lost his security. He misses the companionship and feeling of safety he has when he's with the cow.

Calves worry and fret when taken from their mothers; in their view they are suddenly alone in a world full of predators with no one to protect them. At weaning time they are old enough and big enough to not need milk anymore and are eating grass or hay, but they still want to be with their mothers.

An important part of learning how to care for cattle is learning how to be a good observer, always alert to how an animal feels. As you learn about your cattle as individuals, you will be in tune with their personalities and habits and how they look and act when they are healthy and sassy. Then, if one has a problem, you will recognize it before the animal becomes critically ill. With early detection of illness, you'll have time to help him with the proper care and treatment.

Keep a close watch on calves after you bring them home. If they come down with pneumonia because of stress, it could become serious or even fatal. A calf gives hints that reflect how he feels (see chapter 3). If he doesn't feel well, don't wait to see if he gets better or worse. Some diseases can worsen quickly; a calf may need antibiotics to recover. If you think he might be getting sick, have your veterinarian look at him and give treatment.

Getting Acquainted

When you first bring cattle home, confine them to a corral until they settle down and get used to the idea of being in a new place. If they are calm adults or gentle yearlings, you can turn them out on pasture later that same day if there are still enough hours of daylight for them to check their boundaries and know where fences are. If it's late in the day, however, wait until morning to put them in your pasture, so they won't have to figure out the fences in the dark. If they seem flighty, pacing the corral, leave them there and feed hay morning and evening until they settle down and relax. This might take one day, or several. Calm cattle settle down rather quickly when you bring them some hay, but wild ones may take longer.

As you acquaint yourself with new cattle, give them time and space. Don't try to get close to them at first when you go in the pen with them, especially if they seem flighty and frightened. Use patience and time to win their confidence. Until the animals know you, they may react explosively if they feel cornered, trying to run past you to get away. If you are too close and a frightened animal doesn't have room to get away, he may even snort and charge at you if he thinks you are a threat to his safety — like the wolves his ancestors had to fight off.

When you approach the cattle pen to feed them, let them know you are there. Speak softly. Move slowly. Try not to startle cattle. If a calf's attention is diverted elsewhere and he suddenly sees you, he may run off. Talk softly or hum. A constant stream of quiet chatter or humming can help gentle a wild and insecure calf. Cattle are curious and will be soothed and fascinated; they will start to relax. If they become curious, they lose their fear; you can come closer before they try to run away. Once they learn to associate you with the food you bring, they become more trusting.

When you are in the pen with cattle, walk slowly and don't look directly at them. They'll relax more if you act like you're not paying attention to them. If you come too close too fast and look directly at them, they think of you as a predator. As they relax, walk closer each time you come. Before long, they will come to meet you when you bring feed.

Handling and Gentling Cattle

Some cows and calves are not as timid; they are curious about you from the beginning. You can use this curiosity to advantage. If you are patient and quiet, they will not be alarmed.

Don't press timid cattle. If they won't come to the hay while you are near it, step back. Cattle have a certain space in which they feel secure. This is their flight zone. It's like a big imaginary bubble. If you enter this zone, they become nervous or frightened and run off. As long as you don't violate that space, they feel safe.

Different individuals have different-sized zones. A wild or timid animal has a large one; you can't come very close or he runs away. A gentle or more curious calf has a much smaller flight zone and does not feel threatened until you are quite close. Once your cattle know you well, their flight zone may disappear. Some will let you walk up to them or pet them on the neck. Respect their flight zone when they are still timid, and you will make faster progress in gentling them. Discover the distance you need to keep as you get acquainted. If an animal gets nervous, step back.

Tame cattle like to be petted. They especially enjoy being scratched in places that are hard for them to reach. Most of them like to be scratched under the chin, behind the ears, and at the base of the tail. Do not rub the top of the head or the front of the face. This encourages the animal to butt you. They do this in play to one another, but head shoving can hurt a person.

Meal Ticket

Use feeding time to your advantage. When cattle begin to associate you with food, they lose their fear and walk up to you. It may take a few more days before you can really close in on them or touch them, but they may stand beside you and eat and learn to come when you call. Cattle are good at associating things. If you have a special call for feeding, they will come when they hear it.

The best time to make friends with cattle is when you feed them. They are always interested in food, so they'll want to come when you bring feed. If cattle are timid, bring their hay and put it where they can see you feeding it, then stand quietly close by. Their interest in the hay will diminish their fear of you; they'll come to the feed and thus closer to you, and realize you are not so scary after all. If you stand still, or sit quietly on the fence, they'll start eating, even if they are wary.

As they become less timid, talk to them quietly, hum, or sing a song. Cattle are lulled by a soft voice; it helps put them at ease. They quickly learn the sound of your voice. If other people (strangers to them) come into the pen or pasture, they may become nervous or frightened, but will relax when they hear your voice because they know and trust you.

Don't alarm cattle by having all your friends come look at them right away. Wait until they've adjusted to their new home and settled down a bit. Once they trust people, they won't be so upset when strangers come around. Some cattle figure out things very quickly and soon enjoy being near people, while others take longer to shed their fear.

Don't make the mistake of babying cattle too much. They can be pets, if you wish, but they still must respect you. Remember that cattle are social animals and are accustomed to life in a group; they are always bossing other cattle or being bossed. They'll think of you as one of the herd, once they realize you are not a predator. You must be the dominant herd member, in their eyes; they must accept you as the "boss cow." If they don't, they will try to dominate you and become pushy.

Cattle in a group always have a social order. The bossiest individual is at the top, and got there by being stronger and more aggressive, winning all the fights. The boss cow gets first choice of feed and water; everyone else has to wait in line at the water trough until she finishes drinking. The other members of the herd have their own fights to figure out who is next in the ranking and who can boss whom. The top cow rarely has to defend her title, because all the others have already learned to respect her.

It is natural for cattle to be bossy or aggressive if they become too confident and think you are a subordinate herd member. They will try to dominate you. Don't ever let an animal get away with pushy behavior. It may seem fun when a calf is small, but not fun when he is larger and stronger.

If an animal starts pushing or butting you with its head when you are feeding or petting, reprimand with a swat. Some animals become very bossy and could hurt you without malicious intent; this is their version of play. Calves and yearlings fight when they play, practicing to see who is toughest and preparing for bigger battles when they grow up. A calf that becomes very familiar with you is no longer afraid and will naturally want to "play fight" with you.

Don't let him do this. If you spoil him by letting him do what he pleases, you will regret it later. Don't let him get away with pushing on you with his head, or someday he may hurt you. If he gets too sassy, carry a small stick to rap his nose when he misbehaves. If you don't have a stick when an animal starts shoving on you, twist its ear. A cow or calf's ear is sensitive, and this is something you can always grab if it misbehaves. This will remind him you are still the boss and he won't keep trying to dominate you. You'll have a safer relationship with cattle if it's built on trust and respect. An animal that has no fear of you must respect you as boss or he may become dangerous.

Be Careful and Gentle

Though a calf or cow is not likely to attack a person (unless a cow is trying to protect her young calf), she can still hurt you accidentally, because of size and weight. Always keep an escape route in mind when trying to corner an animal or herd him into a chute; give yourself enough space to be able to dodge aside if the animal backs into you or runs back out of the corner. Don't be in a position with nowhere to go if he suddenly turns and comes your way; he may smash you into the fence.

Cattle can be dangerous when handled in a confined area if they panic. Don't wave your arms, scream, or use a whip when trying to herd cattle. If an animal won't move forward into the catching area or chute, prod him with a blunt stick or twist his tail to make him move. Be careful to not twist it too hard. You can twist it into a loop or push it up to form a sideways S curve. Stand to one side in case he kicks.

Don't yell or chase cattle or you may frighten them. They may run blindly and crash into the fence or you. Even if an animal is suspicious or stubborn and won't go into the catch pen or chute alley the first time, don't become impatient. Losing your temper and yelling will only confuse or scare him. He'll be harder to handle the next time.

Be patient and try again. Keep moving him (or the herd) quietly around the corral until he does go into the catch corner or chute alley. Cattle are more predictable and manageable when they are calm. Speak softly, move slowly, and let the animals move at their own speed. If you are excited and run around, they are more worried about your actions than about moving in the proper direction. Beating and whipping on cattle will destroy their trust.

Feed and Water

When you first bring cattle home, have feed and water waiting for them in the pen. If there is something for them to eat when they arrive, they will be happier. If you are bringing home a group of newly weaned calves, keep them in the corral several days or a week, until they are content with their new situation.

Leave some hay where they can find it easily, but not in a corner or along the fence where they'll be walking on it every time they pace around the corral looking for a way out. Walked-on hay does not taste good; they'll waste a lot of it. Put the feed in the middle of the pen, or along a side they'll be least likely to pace back and forth, or in a feed rack. A group of newly weaned calves are more worried about trying to escape than eating.

For newly weaned calves, feed really good grass hay that is fine and palatable. A calf that is more worried than hungry won't be interested in poor hay; it should be something really good to entice him to eat. Do not feed rich alfalfa, however, or it may make him sick or bloated. Good grass hay or a mix of grass and alfalfa is safer.

Cows or yearlings usually figure out the water if it's in a tank or trough. Newly weaned calves may not have experience with water tanks or tubs if they have been on pasture or range with their mothers, drinking from streams or ponds. Put the water tank or tub next to their feed the first day, so they'll find the water when they come to eat hay. If they soil the tub or tank with feed or manure, rinse it out.

If calves do not figure out where the water is right away, you may have to let a little run over the trough to make a small puddle. If they are used to looking for water on the ground, the puddle (or the smell of water on the ground) will attract them, and then they'll find the tub or tank.

After they learn how to drink out of a tub or trough, move it farther away from the feed area or feed in a different location. If you don't, they may continually spoil their water with feed or manure. If

Cows or yearlings usually drink the water if it's in a tank or trough.

you only have one or two calves and are watering in a tub or bucket, hook it to the fence or situate it off the ground a little ways so they cannot put their feet in it.

Vaccinations and Other Management Procedures

Check with the former owner to find out what vaccinations the cattle have had so you will know whether or not you need to vaccinate them soon or wait awhile before giving their booster shots.

Adult cattle usually need annual or semiannual vaccinations against certain diseases. If you know when they were last vaccinated, you can plan their next shots for the appropriate time. If you do not know the vaccination history on the cattle, it might be wise to vaccinate soon. Ask your veterinarian.

Weaned calves need several shots immediately, unless they were vaccinated just before weaning by the previous owner. Even if they were vaccinated as babies, they need booster shots at weaning time. Consult with your veterinarian as to which vaccines should be given. Have an experienced person do the vaccinating until you learn how to do it.

If you buy calves or yearlings, males should have already been castrated. Most farmers and ranchers castrate bull calves soon after they are born. If by chance a calf is still a bull, have your veterinarian castrate him. If you plan to raise cattle, you will eventually need to be able to castrate your own bull calves (see chapter 10), but then they can be done as small babies. This is easier on them and easier for you.

Castration when a calf is young can be done by putting a rubber ring over his scrotum, above the testicles. When he is bigger, testicles should be removed surgically. A slit is made into the scrotum with a clean, sharp knife, and the testicles are removed.

If cattle have horns, they should be dehorned. Horns can be dangerous; cattle may injure one another or you. Some breeds are polled, but others need to be dehorned. This is usually done when calves are small, but sometimes a calf gets missed or his horns were not obvious when he was a baby and they start to grow later. Dehorning is also easier on the calf when he is small. A big weanling may have horns two or three inches long. These must be cut off and blood vessels tied off or seared with a hot iron in order to prevent the animal from bleeding too much.

If any of the calves you bring home need to be dehorned or castrated, but have just been weaned, wait at least two weeks until they are over the stress of weaning. Have an experienced person or your veterinarian perform these tasks. Dehorning or castration done improperly on a large animal could endanger the life of that animal, due to excessive bleeding or infection.

CHAPTER SEVEN

Natural, Organic, and Grass-Fed Beef

T HERE IS A GROWING NICHE MARKET for several types of beef that are raised a little differently than animals marketed through customary channels. Many people today consider the food they eat as an important part of a healthy lifestyle. Sales for organic, natural, and grass-fed beef are increasing, and many small farmers help supply this market. Some consumers also want to be sure the meat they eat contains no antibiotics or growth hormones, and some want to buy grass-fed (rather than grain-fed) beef. Some also want to make sure the meat they buy is homegrown and not imported, to be free of diseases like BSE (Bovine Spongiform Encephalopathy).

The market for specialty beef is still quite small, however — less than five percent of the total beef industry. Most of this meat is marketed through health food stores or to grocery stores through companies and cooperatives, such as Coleman Natural Meats, Laura's Lean, and Oregon Country Beef, and small regional groups.

Labels on beef in a store can give you a clue as to how it was raised, but terms like grass-fed and natural are not U.S. Department of Agriculture labels. There is no official designation for these terms. However, a producer that labels a product "grass-fed" must have a

USDA-approved system and verification that the animals were grass fed, with affidavits and proof.

The term "organic" is an official USDA certification; animals must have been raised in a certain type of environment, eating feeds that meet particular requirements.

Natural Beef

The term "natural" simply means minimally processed. Almost any beef can be labeled natural, but the term generally means the animal was raised with no antibiotics or growth-stimulating hormones. To market animals through a company like Coleman Natural Beef, for instance, producers must sign affidavit papers to prove they did not use antibiotics or hormones; animals also must be individually identified from birth to slaughter. Any animal treated for illness with antibiotics must be pulled out of the program and marketed a different way after it recovers and reaches market or slaughter age.

Health Care

Since natural beef (and organic beef) is grown without use of antibiotics, the health care and management of the cattle are very important. A clean and healthy environment is always important when raising livestock, but especially crucial for those raised without benefit of antibiotics. Cattle in a natural beef program need lots of room and should never be confined in a dirty area. Diseases like calf scours, coccidiosis and other problems that occur when animals are confined in close quarters can wipe out an attempt to raise animals without medication. Avoid these problems by keeping animals in a very clean environment.

Feed Program

Natural beef can be either grass-finished or grain-finished. Colemans Natural Beef, Laura's Lean, Oregon Country Beef, and several other branded products use a finishing time on grain. The length of time on grain varies with the goals of the specific program.

The feeding process for natural beef is basically the same as with any cattle feeding program except the diets are verified; the feedlots must show proof that there were no ionophores used. Ionophores like monensin and lasalocid are used in feedlot cattle to help prevent bloat as well as to increase feed efficiency. Their antibiotic action helps prevent or reduce the problems like acidosis, bloat, stomach ulcers, and liver abscesses that often occur with a high-grain diet; and cattle gain weight faster and more safely. Since these compounds cannot be used in the production of natural beef, more roughage is included in the ration (such as ground corn cobs), to reduce the risk for bloat. Adding roughage is similar to the way cattle were fed 50 years ago. Since cattle can't grow quite as fast on this type of program because they don't receive growth implants and ionophores, they may be in the feedlot longer to reach the desired endpoint. The cost of gain is about 10 to 15 percent higher than with typical feedlot cattle.

The breeds or crosses that work best for natural beef depend in part on your situation and goals. Since growth hormones can't be used, fast-gaining continental (exotic) breeds or crosses are often used, in order to achieve a large carcass at a young age. The animals are butchered at a relatively young age, when meat is still quite tender. Those in a grain-feeding program are put in the feedlot early — such as right after weaning — instead of being "backgrounded" on pasture or cornstalks.

Organic Beef

The one official USDA label on natural beef is "organic." Meat marketed under this label has been raised and fed in a program using a recognized national standard, and all USDA specifications have been followed. "Organic" means there were no human-made chemicals used at all in the production of that meat, although vaccination is allowed. It also means the animals were raised in a humane and natural fashion, on land properly managed for good environmental health and sustainable food production. There is no overgrazing or damage to the land. The producer must write an organic system plan and allow official examination of the program. Consumers want to know that these animals were raised in a manner they find acceptable.

Organic Feed

Beef raised in an organic meat program must be fed naturally-grown feed, whether it is grass, hay, or grain. No pesticides, herbicides, or chemical fertilizers can be used on the feed crops. Feed purchased for the animals must be bought from a certified grower who raises only organic crops.

There are many farmers and ranchers who raise hay and grain crops naturally, but their crops still do not qualify as something that could be fed to organically grown beef animals because the crops are not certified as organic. Since it is often difficult to find organically grown hay and grain, some farmers who market organic beef complete the entire process themselves — raising everything that is fed to their animals — to ensure the feeds are truly organic. Some raise grass-fed beef, growing the pasture and hay themselves, and some finish their animals on grain that they grow themselves.

Organic beef is primarily defined by three things: the producer can verify where the animal has been, what it has been eating, and that it has never had contact with human-made chemicals.

Chemical Free

An animal that will be processed as organic beef can't graze in an area where herbicides or pesticides were used. Even if you do not use these on your own land, the animals cannot have contact with a roadway where the county has used a weed spray, for instance. They cannot graze a pasture where apple trees were sprayed to control damaging insects. They can't graze cornfield aftermath where herbicides or pesticides were used.

Some of the standard cattle treatments and preventative health-care management programs are not allowed when raising organic beef. Organic beef producers can't use pour-on insecticides to control flies, lice, grubs, or other parasites, nor can they use fly tags or de-worming drugs. Organic standards in the United States are some of the toughest in the world because of the total prohibition of chemical dewormers, which are allowed in European and New Zealand organic regulations. Many farm pastures that are used for raising cattle because of their lush, green growth (good moisture conditions from rainfall or irrigation) are also most risky for the spread of worms.

The organic program does allow vaccination, however, since some diseases must be prevented or animals can die. It is acceptable to vaccinate all cattle against IBR, BVD and some of the other viral diseases, and clostridial diseases like blackleg, malignant edema, and redwater. It is essential to vaccinate cattle against the important diseases in your geographic area. This means every calf must be vaccinated, and all cattle vaccinated once or twice a year (see chapter 3). Proper vaccination is essential to cattle health.

The stipulation against antibiotic use does not mean a producer must let an animal suffer or die if it becomes ill. The occasional sick one can be treated, but it must be pulled out of the organic program after treatment and marketed a different way; it cannot be sold through a natural or organic beef program.

Part of the philosophy behind raising organic food is that methods used must be healthy for the animals and for the environment. Cattle are raised on pastures rather than confined in a feedlot. Even animals finished on grain are often in a large pen or small pasture, with room for exercise. They are not putting a lot of impact on the land. The producers of organic beef say they are being better stewards of the land and animals because they are raising cattle a better way — the way they used to be raised, before modern technology and mass production of feed grains and feedlot beef.

Proof of Process

Producers who sell their meat through an organic or natural program, or a health food store, usually fill out an extensive affidavit stating how they raised the animals and whether they were given vaccinations. They must specify the vaccines and how often they were given. Antibiotic use — when and how often — must be documented. They may have to state whether they used certain fertilizers or pesticides. The affidavit must also state how far the animals were transported for harvest (to the slaughtering facility).

Many health food stores purchase meat from small family farms or from cooperatives made up of farmers and ranchers who raise their animals a certain way. The customers know they have affidavits from the farmer or rancher on how the animals were raised, or have a direct relationship with the producer. This is a big selling point; the customers like to know where the meat comes from. The animals are identifiable.

Customers often ask about the meat, and request telephone numbers of the people who raised it. The customer can call the farmer and ask questions. Many customers like to talk to the person who raises the animal they are going to eat. They know the animals were raised from conception to slaughter by a person who cared about them and took good care of them.

At a regular grocery store or restaurant, customers usually don't know where the meat came from, although some restaurants may have a specific, steady supplier. Many consumers want to know exactly what they are buying. The natural beef market will continue

Customers often like to know where their meat comes from, to make sure the animals are treated properly.

to grow as people become conscious of food safety issues and become aware of how animals are raised. They want to buy meat from animals that they can personally identify with, and know that they were raised with humane and conscientious management practices.

People who buy natural, organic, or grass-fed beef from small family farms usually don't mind paying extra for it, if it comes with the knowledge of where the beef came from and how it was grown. Customers who visit with the producer and talk about the meat offer feedback that can help a beef producer fine-tune the way he feeds and processes animals to improve tenderness and flavor. Meat from a grass-fed animal, for instance, may not always be tender, depending on how old it is, the time of year it was slaughtered, which determines whether it was on lush, green pasture or on hay, and what it was being fed. It's not always easy to produce perfect beef under totally natural conditions!

Grass-Fed Beef

Natural and organic beef can be grass fed or grain finished. Some producers are opting to finish them on grass, feeling this is more natural and healthy for the animals and for people eating the meat. Some producers compromise, raising cattle primarily on grass and then finishing for a very short time on grain. The beef is still lean and has the health benefits of a grass-fed animal, but can be brought to market quicker and in better butchering condition during the off-season (winter) if it finishes on grain.

High-quality Forage

If you don't have grass year-round or if grass is poor during drought, the animals must be fed high-quality hay. Since grass-fed cattle are generally raised without supplements, the hay must contain adequate amounts of protein, vitamins, and minerals. Whether you grow hay or buy it, make sure it was harvested under ideal conditions. It should be cut while still growing and tender, and not coarse and mature. Hay that is harvested at an immature stage has more protein and vitamin A, for instance, than hay cut after it has bloomed or gone to seed.

Grass-finished beef has more variation in flavor than grain-fed beef because of a less consistent diet. The best flavor in grass-finished animals comes from using pastures with no weeds like wild onion or other plants that influence meat flavor. One reason some of the natural beef programs put cattle into a feedlot for a short finishing time is to assure more consistently flavored meat.

One of the better winter forages for a year-round grass fed beef or dairy program is annual ryegrass. Compared with other winter annuals such as cereal rye, triticale, or wheat, ryegrass is the most nutritionally balanced and also has a longer productive growing season. It does not decline in quality as quickly.

If you live in a region like the Southeast with season-long grass, raising grass-finished beef is a way to keep cattle until they are ready for slaughter. You can be less dependent on grain markets and the fluctuating prices for feeder cattle.

Raising animals without grain means a longer feeding program to achieve finish size. There is more money, labor, and feed in each animal (feeding hay when pasture is not ideal), and producers must wait longer to see a return on their money.

Meeting the Quality Challenges

There can be a vast variety in flavor and tenderness with grass-fed animals, depending partly upon their age, the time of year they are butchered, and body condition. One reason most cattle in this country are finished with grain is that they gain weight faster on a grain diet; it shortens the time required to raise them to finish size. The animals are young enough to still be very tender. Grass-fed carcasses must be aged longer (before cutting and packaging) to be as tender as grain-fed beef. A grass-fed carcass should be dry aged for at least 14 to 21 days.

Grain feeding also helps create a product to sell year round, because cattle can be finished even in the dead of winter when there is no grass. This way farmers aren't dependent on seasonal marketing. Grain-fed beef is often more consistent in quality and flavor. Grass-fed animals can compete, but only if producers are careful and diligent in their feeding and marketing methods.

Grass-fed beef may be inconsistent in flavor and tenderness because of differences in feed (types of grasses) and hay supplies. Even weather can make a difference from year to year, determining whether grass grows fast or slow. Grass that grows slowly usually has more nutrients per volume than fast-growing grasses. Mountain grasses grown at high elevations may be more nutritious than grasses that grow more swiftly at lower elevations, but the grass growing season is also a lot shorter in the mountains. Soils can also be a factor in the nutrient and mineral quality of forage plants.

The difficulty in trying to furnish grass-fed beef to a specific market is in supplying a product year-round. Cattle finished on grass are usually best if calved at times of year to take best advantage of grass seasons, finishing as long yearlings or two-year-olds in summer when grass is at its peak. Pastures should be closely managed to use the best forages for rapid gain before the animals are marketed. For cattle that are harvested in winter, it's more difficult to provide a forage that will optimize meat quality. The hay must be exceptionally good. For the small producer, it's often easiest to sell animals locally, butchering when they are most ready and aiming for a summer finish time.

Whether you are part of a cooperative that markets beef for stores or you market it to health food stores yourself, the beef must be slaughtered and packaged at a plant that is USDA inspected. About 20 percent of the cost of raising beef for this market is in the processing — transporting animals to the packing plant, getting meat cut and wrapped in vacuum packaging, and distributing it to the stores. If you only raise a few animals, it is simpler to sell direct to the consumer. Customers who want a whole or half beef for the freezer can purchase it by the head or by the pound and then have the animal processed by a custom butcher who will cut and wrap it to the customer's specifications.

Health Benefits

The meat from leaner, grass-finished animals is not heavily marbled and does not have as much extra fat to be cut off the outside. Cattle fatten on grass, but it's a different kind of fat. It has more omega-3 fatty acids (the good kinds that are necessary for good health) and less of the omega-6 fatty acids (the "bad" ones, that are more prevalent in corn-fed beef). Omega-3 comes from leafy plants like grass, whereas the omega-6 fatty acids come from grain.

One obstacle to consumer acceptance is the yellowish fat of grass finished beef, rather than bright white fat of grain-finished beef. The

Yearling steers grazing on lush pasture.

yellow color is caused by the higher beta carotene (vitamin A) content of green grass.

Grass-fed beef (and milk from cows grazed at pasture rather than a high level of grain) contains more conjugated linoleic acid (CLA), a fatty acid found in beef and dairy fats. The human body does not produce its own CLA, but can obtain it through foods like whole milk, butter, and beef. CLA helps promote healthy cell structure and regeneration. It has been shown to help reduce risks for cancer and disease. Grass-finished beef is lower in total fat and calories than grain-fed meat. It is a more natural diet for humans; we ate grass fed animals for thousands of years.

Some natural beef producers utilize a short finishing time on grain, to help create more consistency in flavor and marbling. The amount of grain fed is minimal.

Traditional feedlot finishing of beef requires a lot of grain. The animals are usually fed low-level antibiotics like Rumensin to help avoid bloat (by reducing fermentation) and prevent liver abscesses and stomach ulcers — problems that can occur in animals on a straight grain ration toward the end of the feeding period. This can be avoided if cattle are finished on grain for a very short time (60 to 90 days versus 120 to 150 days) and the grain levels are not high.

Animals fed short-term on grain do not lose the benefits of the grass diet. The meat still contains a good balance of fatty acids with higher levels of omega-3 from the lifelong grass diet that led up to the short feedlot experience, and lower levels of omega-6. This is often a good compromise for people who want to produce natural or organic beef in a region without year-round grass, or to create a product with more consistency.

Producers interested in marketing natural, organic, or grass-fed beef can check the Web site put together by Colorado State University and American Farmland Trust (see Helpful Sources, page 253). It supplies in-depth information on various aspects of niche beef marketing. The Helpful Sources section also provides an example of a cooperative that markets natural beef.

CHAPTER EIGHT

Raising Beef Heifers

I F YOU WANT TO RAISE CALVES you don't plan to butcher or sell when they grow up, choose heifers. You can start a herd of cattle, selecting a group of weaned heifer calves or yearling heifers to be the foundation of your herd.

Selecting Beef Heifers

Selecting heifers is a little different than selecting steers to fatten for butcher or market. The beef steer is judged on beef characteristics — lots of muscle and good frame. For beef steers bound for the table, you want animals that will grow fast and put on weight in a short time. A breeding heifer should also be able to grow fast. Even more important than ability to put on weight is her ability to become a mother cow. The biggest, fattest heifer doesn't always make the best cow; her body is programmed more for getting fat than for getting pregnant or giving lots of milk for her calf. She may have too much fat in her udder, displacing mammary tissue, and she may not milk well. A big, muscular heifer that lacks feminine characteristics may not be as fertile as a more slender, moderate-framed heifer.

What Breed?

The breed you choose depends on personal preference, breeds available in your area, and whether you want a registered purebred, straightbred commercial heifer, or a crossbred or composite.

A registered purebred may cost more than a commercial heifer. This does not mean she is better; it means she has registration papers. You should receive those papers when you buy her. You can sell her calves as registered stock if you breed her to a registered bull of her breed. To sell her calves as purebreds, you must register them when they are born. To raise registered cattle, you must join the breed association and pay registration fees for every calf you register.

Most breed associations have regional managers who work with cattle breeders to promote their breed, help people get started in the purebred business, and help them improve their herds. Contact the association and someone will direct you to local breeders who might sell you some heifers. Once you buy your registered heifers, give the seller all the information needed to transfer their registrations to you.

Registration papers give the pedigree and registration number of the registered animal, certifying that this animal is recorded in the herd book of its breed.

A straightbred is an animal with parents of the same breed, but not necessarily purebred. There may not be much difference between a straightbred and a purebred, except that the ancestry of the straightbred is not verifiable. A straightbred will usually be less expensive than a purebred but it will not have records, which may not be important to you if you are just raising commercial cattle.

If you just want to raise good cattle that produce top-quality market calves, they don't have to be purebred. The best beef cows are usually crossbreds or composites, with the "plus" of hybrid vigor (see chapter 4, page 72). Just make sure the heifers you select have good conformation and feminine characteristics.

Feminine Features

Heifers that look like steers are not good choices as breeding stock. Good cows look feminine rather than masculine. Picture in your mind what a good dairy cow looks like and you'll understand what is meant by feminine traits: a relatively slim, angular body (in contrast with the bulging muscles of a bull), a fairly long, narrow head and neck (graceful head and neck rather than short, thick bull neck), and a well-developed udder.

The beef cow is not as extreme in these feminine traits as a dairy cow. She should have good muscling and a smaller udder. She still should look graceful and feminine. Big, muscular, or overly fat heifers that look like steers are often not as fertile as they should be and may not have a calf every year. A heifer that is too fat from the time she was small has too much fat deposited in her developing udder, which hinders milk production later when she needs to feed her calf.

Conformation (body structure) and disposition are two things to evaluate carefully when selecting heifers. Performance records cannot measure these qualities.

Performance records

If you buy purebred heifers, use the breed's performance records and expected progeny differences (EPDs) to help in the selection. Successful breeders keep detailed records and use them to identify genetic differences in the cattle. You can use this information to compare things like birth weight, weaning and yearling weights, milk production and fertility.

Progeny means offspring. Using EPDs, cattle are compared with one another. Numbers for weaning weight and other variables are put into a computer to come up with a score to show how they rank in the herd, or in the breed, on certain traits. When selecting a heifer as a future cow, you are also selecting her contribution to the possible genetic traits of all her offspring. The EPD is a tool for helping predict the possible characteristics of her calves.

EPDs are an estimate of an animal's genetic ability to transmit a particular trait, compared to other animals in the breed. A maternal (milking) EPD on the dam, for instance, shows how the heifer's mother compared to other cows of the breed. A maternal EPD on her sire tells how well his daughters milk. EPDs are based on averages.

Herd average for a specific trait would be zero. A cow with a plus number indicates her milking ability is that much above average; a cow with a minus number means she's that much below average.

EPDs can be confusing because breed average for a certain trait may no longer be zero. A breed EPD for birth weight or weaning weight may have averaged zero when the evaluation system was established, but as more cattle are evaluated every year, trait averages change. In many British breeds, average cattle are bigger now than when the system started. A plus number on weaning weight may actually be average today. Ask a breeder to help explain EPDs to you and how to use them in selecting cows or heifers.

Performance records rank each animal to see how it compares to other individuals in the herd or in the breed. The heifers you are considering should have a recorded birthweight, weaning weight, and yearling weight, if they are old enough. You can look at records for their sires and dams. When selecting purebred heifers, have a breeder explain the records, so you can understand how to use them in comparing animals you are considering.

An important thing to check is birth weight. Choose heifers that were small at birth (average or lower birth weight), yet grew fast to have a respectable weaning weight. Heifers that were large at birth may have calves that are large at birth. Big calves can result in difficult calving, which is something you want to avoid.

Unregistered heifers won't have performance records to help you evaluate their potential. The stockman who raised them can often answer questions about the heifers' genetics. He can usually tell you about the bull that sired them, such as whether his calves are born easily and how big the calves are at weaning. He might also be able to tell you something about the heifers' mothers. If he uses individual identification (ear tags on the calves correspond with the mother's tag number) he might tell you about the mother of each heifer, such as whether she's a good mother, or calves easily, how may calves she's had, whether she calves early or late in the calving season, and whether she has a good udder. These are very important questions to ask, whether the cows are purebreds or commercial.

When buying crossbred or composite heifers, a good mix of genetics from different breeds will usually produce an outstanding individual that may outperform most purebreds in all traits because of hybrid vigor. A heifer must have good parents to be able to do this. Just because she is a crossbred does not mean she will automatically be a good cow. She must have good genetics from both parents. Find out as much as you can about her sire and dam and evaluate her body structure, disposition, and udder shape.

Pedigrees are helpful, when available. A pedigree is a chart showing the sire and dam of the calf, and their parents and ancestors. The sire is the father; the dam is the mother.

Conformation and Frame Size

Conformation refers to how the animal is put together and is important when selecting heifers. Cattlemen judge a cow as much by how she looks (the way she is built) as by any other factor. Her structure is as important as her performance records. Heifers must have good feet and legs and a good udder. Even if a heifer has good performance records and is genetically programmed to wean big calves, she will disappoint you if she becomes crippled early in life and must be culled (sold) from the herd because of poorly built feet and legs. She will also be a cull if she has a poor udder that breaks down after only two or three calves.

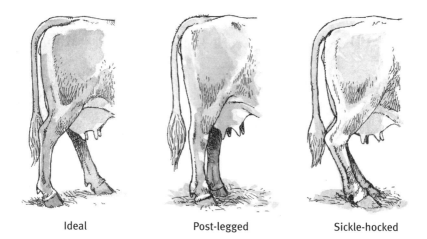

Ideal   Post-legged   Sickle-hocked

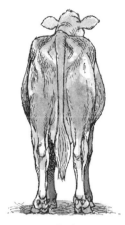
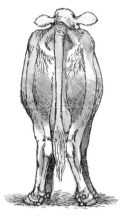

| Ideal | Cow-hocked | Feet too close together |

A good heifer must have a long body. You want her calves to be long; they will be better beef animals, with more capacity for putting meat on that frame. A long body also gives more room for carrying a calf during pregnancy. She should have a deep body, not shallow or narrow. She should have good muscling, though not excessive. She should move freely as she walks, with nice athletic ability. She should have that elusive quality called style — all her parts fitting together to create a good-looking animal.

Choose heifers with medium frame size. A small cow will not raise big calves. A big heifer may grow into a huge cow. It may take her too long to reach breeding age and she also may not be as fertile as a moderate-framed animal. Cows that are too large are not efficient. They require too much feed for what they produce. A medium-sized cow is more productive because she generally weans a calf that is larger, in relationship to her own body size.

Moderate sized cows are most profitable. A 1,400-pound cow that weans a 700-pound calf (only half her body size) has not done as well as a 1,000-pound cow that weans a 600-pound calf (more than half her own weight); the smaller cow ate a lot less feed than the 1,400-pound cow. The 600-pound calf cost less to produce and thus made you more money than the 700-pound calf. You could feed a larger number of small to medium-sized cows on the same feed and pasture, and have more calves to sell.

You want heifers that grow fast and reach mature size quickly for early puberty (sexual maturity), so they can breed as 15-month-old yearlings and calve as two year olds. Larger framed heifers often take longer to reach their mature weight and reach sexual maturity later. They are generally not as fertile or as productive during their life as moderate-framed females.

Udder

Milking ability and udder shape are important factors in determining whether a heifer will be a good cow. You want her to give lots of milk when raising calves, so they will grow fast. She must also have a well-structured udder. You don't know what her genetic capacity for milking ability is unless you know something about her sire and dam (and the potential she inherits from them), but you can carefully evaluate her udder. You don't want a cow with fat teats, long teats (which a newborn calf might have trouble getting into his mouth), or a saggy udder that might become injured. A cow does not need a big, droopy udder to give plenty of milk.

You can't tell by looking at a heifer's udder exactly what it will be like when she matures, but there are clues. Select heifers with nice small teats of uniform size and length. Fat, long, or uneven teats (front teats longer or shorter than rear ones) at this young age will become even worse when she grows up. Poor udders are a serious fault; many beef cows have to be culled because of this problem.

Udder shape and size are inherited. When selecting a heifer, it helps to know about her sire and dam. Ask to see her mother. Sometimes you can look at heifers before they are weaned and evaluate

teats too fat   teats too long   unbalanced quarters

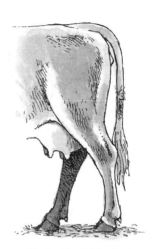

Ideal udder (back view)   Ideal udder (side view)

their mothers to see what the heifers will grow up to look like. Ask about the calf's sire and what kind of udder his mother had, too. The bull has as much or more influence on a heifer's udder as her mother does; females often inherit traits from their father, so udder traits can come from the bull's mother. For that reason, when a stockman selects a new bull, he usually tries to look at the bull's mother. Heifers he keeps from a bull will be a lot like the bull's mother in looks, conformation, and udder shape and size. If possible, look at young cows in the herd with the same sire as the heifers you want to buy.

Disposition

A cow's disposition is partly genetic. She inherits from her parents a tendency toward being nervous or placid, flighty or calm, smart or stupid, kind or mean. Just like humans, some cattle are smarter than others, and some are more emotional. Some breeds tend to be more flighty than others, but there are "wild" and "calm" individuals in every breed. Do not assume that a heifer will be easy to handle just because she is a certain breed. Disposition is also influenced by how the heifer is handled and managed. A timid, nervous heifer that is smart will often gentle down with patient handling and learn to trust you. She's smart enough to realize you are not going to hurt her.

On the other hand, some wild and nervous animals never do figure that out and can be frustrating and dangerous to handle. A gentle cow can become wild and untrusting if she is handled roughly or with shouting and whips.

A heifer's attitude is especially important if you plan to keep her as a cow. You want smart, gentle cows that will be a pleasure to handle. You want to be able to work with them without danger of being hurt. A wild or mean cow may become very excited and aggressive when she calves; it will be difficult or dangerous to do anything with her or the calf. You want good-natured cows that pass on their gentle attitude to their calves.

When choosing heifers, pay close attention to how they act. Look at their mothers; this can also be helpful. If the mother is wild and suspicious, or gentle and curious, the heifer will probably be a lot like her, unless the heifer inherits a very different nature from her sire.

Care of Beef Heifers

Care for heifers is like any other beef calves except you'll feed them differently than market beef animals. Proper nutrition is important. Feed heifers for optimum growth, so they will reach puberty (see chapter 9) soon enough to be bred on schedule during their first breeding season. Do not overfeed them to the point of fatness, since this can be detrimental to fertility and future production.

Diet

Good pasture during summer is ideal feed for a growing heifer. This can also save money because you won't have to buy hay and grain to feed her. Green grass provides all of her nutritional requirements except for salt and water.

You shouldn't have to feed heifers much grain, if any. Heifers that need grain to grow fast enough to meet ideal breeding weight and mature weight on schedule won't be good cows. You want a heifer that will grow well and produce good calves without pampering. Feeding grain can defeat your purpose in getting her to reach her target breeding weight: She may become fat and reach that weight before she is sexually mature enough to breed. She has not grown enough in frame, so the extra weight is just fat.

Fat is not healthy for a breeding heifer. Too much fat around her reproductive organs makes her less fertile. She may not breed as quickly as a leaner heifer and might not become pregnant when bred. If she is too fat during pregnancy, she may have trouble calving. It will be harder for her to push the calf out and fat deposits around the pelvic area and birth canal may hinder progress of the calf as it comes out. Fat is also detrimental to milking ability. When too much fat is deposited in the udder, a heifer doesn't milk well when she has calves. Fat displaces the developing mammary tissue (the milk-producing glands), and she will never reach her genetic potential for milking ability. Cows live longer and stay healthier when they are well fed but not overfed.

Some producers feed heifers grain so they will gain weight faster, especially in purebred herds. Some breeders believe grain makes bulls and heifers look better in the sales or at shows. Purebreds usually sell for more money than commercial cattle, so breeders may feel they can afford to use grain or buy extra feed to make young cattle grow bigger and fatter more quickly. Although the grain-fed young bull or heifer looks good, this is not a true indication of how that animal would perform on just grass.

Most commercial cattlemen who depend on what their cows will produce without pampering and extra feed expense need cattle that can do well in the "real world" without pampering. Many stockmen raise crossbred cattle that grow fast on hay and pasture or on the sparse bunchgrasses of desert rangelands or high mountains. If a heifer must have grain or expensive supplements to grow big enough fast enough to breed and calve at target size and weight, this takes the profit away from the calf she will produce. The best kind of cattle to raise are efficient ones that grow fast and milk well on natural feeds, raising big calves and still breeding back again on schedule, eating whatever feeds the farm or ranch can produce cheaply, such as grass and hay.

When raising heifers to start a cow herd, choose heifers that will grow nicely and breed quickly without grain. Most good crossbreds will do this, as will some purebreds; you just need to be careful in your selection when picking heifers to buy.

A clue to heifers' genetic ability for feed efficiency is the feeding management in the herd where they originated. Ask the rancher or purebred breeder how he feeds his heifers — whether he grows them on just forages, grass, and hay, or feeds them grain.

Rate of Growth

Heifers need good feed to grow big enough to reach puberty at an early age and be ready to breed by the time they are 15 months old. A heifer needs to have 65 percent of her mature size by the time she is about 14 months of age.

The actual weight she should be at this age will depend on her breed and frame size. Angus cattle of moderate frame size tend to mature at a lighter weight and so are ready to breed at an earlier age than Herefords or Simmentals, for instance. A moderate-framed Angus heifer that weighs 650 pounds at 14 months will be just as ready to breed as a Simmental heifer that weighs 800 pounds, and it may take the Simmental a little longer to reach 800 pounds; she may be 16 to 17 months old by that time. Know what your heifers should weigh for their age, breed, and frame size, and feed them accordingly.

Some people raising breeds that were traditionally moderate in size, like Angus, have been selecting for larger-framed animals to try to compete with the bigger European cattle on weaning weights, so you can no longer make generalizations regarding weight at puberty. There are Angus cows now, for instance, that mature at 1,200 to 1,400 pounds. These big-framed cows will be larger before they reach puberty than will moderate-framed traditional size Angus cows. They are now more like the larger-framed breeds and also have larger calves at birth. A large-framed Angus may be slow to reach puberty and may also have big calves that are *not* born easily. You must take frame size into consideration when predicting weight at puberty and readiness to breed.

To be of proper size and maturity for breeding, British breed heifers should weigh at least 650 to 700 pounds (depending on frame size; larger framed animals will be at the high end of that scale or even higher) and European breeds should weigh as much as 800 to 850 pounds.

If you want heifers to calve in mid-March, they must be bred in early June of the previous year; pregnancy takes about nine months plus seven days. If today is March 3, you have about three months until the heifers should be bred. If they are moderate-framed Angus and weigh about 550 pounds today, they need to gain about 100 pounds to reach target breeding weight of 650 pounds. They should do that easily, gaining a little more than a pound per day for the next 90 days. Heifers can do that on a good roughage diet such as a mix of good grass and alfalfa hay until pasture is ready to graze, and finish up the time on good pasture. They should not need grain.

Sometimes good pasture is not available. During a drought, the pasture grasses dry up and become less nutritious or don't grow well. Maybe you don't have enough pasture area for grazing all summer and fall. There might be times you have to feed hay or grain. If pasture is dry — low on protein and other important nutrients — you can supplement with alfalfa hay or some really good grass hay, if available.

When hay supplies are expensive or poor quality because of drought, you may find a little grain and a protein supplement to be a better buy. Consult your county Extension agent, a cattle nutritionist, or a successful farmer or rancher to help you figure out a good ration to keep heifers growing and able to breed on schedule without losing weight or becoming too fat.

With proper feeding management, weaned heifers can continue to grow at normal rate on poor pasture and a grain/protein supplement. If a heifer weighs between 500 and 600 pounds at weaning and has genetic potential to weigh 750 to 800 pounds at breeding age (15 months), without being fat, she must gain 200 to 300 pounds in the 160 days between weaning and breeding age. She should do this easily on good hay over winter and pasture grass in the spring, especially if she has the genetics to be feed efficient and a good gainer. If you decide to add grain, start heifers on just a small amount at first and increase it gradually to avoid problems. Never overfeed on grain.

Monitor how your growing heifers look. If they are not growing fast enough or seem a little thin, supplement their pasture. If it's winter and you are feeding hay, feed them more. A good rule of thumb when feeding hay is to feed them all the good grass or grass/alfalfa hay they will clean up.

You really cannot overfeed a growing heifer on non-alfalfa hay. Rich alfalfa and grain must be rationed to avoid bloat or founder. Monitor heifers for too much gain to avoid fat problems. Heifers that get too much feed convert the extra calories into fat. If heifers start to become fat, cut back or eliminate the grain. Carefully evaluate their growth and body condition and adjust the feed accordingly.

Keep in mind that as heifers grow and become larger, they need more feed. The amount of hay and/or grain that was adequate when they were 600-pound weanlings will not be enough as they become 800-pound yearlings. Gradually increase the amount of feed as the heifers grow.

CHAPTER NINE

Breeding and Calving the Beef Heifer

G ESTATION LENGTH FOR CATTLE IS NINE MONTHS plus a few days. If you want to calve in March, breed in June. If you want to calve in May, breed in August. When breeding a group of heifers, make sure they are at least 14 to 15 months old at time of breeding and have grown enough to reach target breeding weight. They must be in good body condition for peak fertility — neither too thin nor too fat.

When to Breed Heifers

Ideal breeding age for heifers is between 14 and 16 months of age. At that stage of maturity, they should be about 65 percent of their mature weight. British breeds need to weigh at least 650 to 700 pounds at breeding; heifers of larger framed breeds need to weigh more. If a heifer is a small-framed Angus that will mature at 1,000 pounds, she must weigh 650 by breeding time. If she is a Simmental-Hereford cross that will weigh 1,300 pounds at maturity, she needs to weigh almost 850 pounds by breeding.

Don't breed heifers too young. Even if they reach puberty early and are cycling, don't breed them before they are 14 months old.

A heifer can become pregnant any time after she is sexually mature and having regular heat cycles, but breeding her too young may result in difficult calving and trouble breeding back on schedule. She may not have reached ideal breeding weight and will be too small at calving. If she's still trying to grow while raising a calf, she may not breed back again on schedule.

For best production, heifers should be bred as yearlings to calve as two year olds. Heifers of most breeds will easily do this, if adequately fed. Heifers that develop slower than this timetable are not as profitable. A heifer that is not sexually mature enough to breed until she's two years old will not calve until she is three. You'll have a lot of feed invested in her by the time she has her first calf, and she'll have fewer total calves during her life.

Most heifers reach puberty by the time they are 12 months old. Some will cycle earlier, and some start later if they are slow-maturing individuals. Some will be cycling by 9 or 10 months old, but are much too young for a healthy pregnancy. You don't want a heifer to calve before she is 24 months old. If she's much younger than that she can't do a good job of raising a calf. If she's much older, she may calve late every year for the rest of her life.

Check the accompanying chart for estimated target weights for some of the breeds and their crosses to see the effect size and weight have on puberty. As you can see from the chart, 95 percent of Angus heifers have reached puberty and are cycling by the time they are 700 pounds, whereas only 65 percent of Charolais cross heifers are cycling at 700 pounds. Most Charolais cross heifers weigh about 800 pounds before they reach puberty, and a purebred Charolais heifer must be even larger before she begins cycling.

Breeding Options

If you have only one or two cows or heifers you might take them to a neighbor's farm to be bred (leaving them there until they come into heat and are bred), or have them bred by artificial insemination (AI).

If you have a small herd, you might consider borrowing or leasing a bull. It is usually not economical to buy a bull unless you have at least 15 to 20 females to breed. Unless you have good fences and a strong pen to keep the bull during the times of year he is not with cows, it's easier to not have a bull. If you lease or borrow a bull, or take heifers to be bred to a neighbor's bull, make sure the bull has had proper vaccinations and tests negative for certain venereal diseases if testing is required in your state. Check with your vet on state regulations.

WEIGHT OF DIFFERENT BREED CROSSES AT PUBERTY

(Heifers are assumed to be at least 13 months old.
Crosses are from Angus or Hereford cows.)

Breed	AT 600 POUNDS PERCENT CYCLING	AT 700 POUNDS PERCENT CYCLING	AT 800 POUNDS PERCENT CYCLING
Angus	70	95	100
Angus/Hereford X	45	90	100
Charolais X	10	65	95
Chianina X	10	50	90
Gelbvieh X	30	85	95
Hereford	35	75	95
Limousin X	30	85	90
Maine Anjou X	15	60	95
Shorthorn	75	95	100
Simmental X	25	80	95
Tarentaise X	40	90	100

Signs of Heat

A cow or heifer can only be bred when she comes into heat (estrus). She must be bred at the proper time to become pregnant. If she is to be bred artificially, you must be able to determine when she comes into heat and then have an AI technician insert a capsule of semen into her uterus at the proper time.

Determining when a cow or heifer is in heat can be difficult with only one animal and no other cattle around. There are some outward signs of heat; the cow or heifer will be more restless than usual and may pace the fence or bawl. She may have a clear or slightly bloody mucus discharge from her vulva. Not all females show obvious signs.

If she's living with other cattle, it is easier to tell when she comes into heat. Other cattle will mount her or she may try to mount them. They may fight more than usual. In a herd of cattle, when there is a lot of activity going on due to one of them being in heat, you can tell which one is in heat; she will be standing still for the others to mount her. This is called "standing heat," when she is receptive to being mounted and bred. The hair over her tail and hips may be ruffled from this activity — another clue that she is (or has been) in heat. If you don't have other cattle with her to help you tell when she's in heat, it may be simplest to take her to live with a bull for one to three weeks until she is bred.

Be careful around a heifer in heat. When cattle are in heat, they goof around and "ride" one another, pretending to mate. A pet cow

A cow in heat may be mounted by other cows.

or heifer that has no fear of you can be dangerous when in heat; she may treat you like she would another cow. People have been injured by pet cows trying to rear up and mount them. Be alert to your heifer's mood and behavior. Don't let her begin any playful actions that could hurt you. Carry a stick to reprimand her if needed, especially when she is in heat.

Selecting a Bull

If cows or heifers are registered purebreds and you want calves you can register, they must be bred to a registered bull of the same breed. If they are crossbred, or you want to raise crossbred calves, choose a bull of a different breed, or use a crossbred or composite bull. If you don't have a bull yourself, you can have them bred by AI.

You might ask the local cattleman who sold your heifers to you if he'd consider putting them with a bull at his place, and what he would charge. Make sure the bull is not the heifer's own sire. Most breeders will be glad to help you. They would probably put your heifers with a group of their own heifers being bred, in a pasture with a bull that sires calves with low birthweight — to help avoid calving problems. If you have only one heifer, he might put her in a pen next to other cattle so he can observe her for signs of heat, then put a bull with her. Ask the bull's owner to keep track of the breeding date so you can predict roughly when the calf would be born the next spring.

Once a cow or heifer is bred, the fertilized egg will start growing in her uterus. Gestation length is about 285 days, but she may calve as much as nine days before or after her "due date." Most will calve within three or four days of their due dates. Some breeds (and family lines within breeds) have a slightly shorter or longer than average gestation length. This is one factor in whether a calf will be small or large at birth. Animals that have low birthweights are usually those with shorter gestation lengths.

When selecting a bull, evaluate all his records in order to know whether his genetics will complement those of your females to produce good calves. One of the main things to take into consideration, however, is birthweight of calves he sires. If he's a young bull that has never sired any calves, look up his birthweight. His breed EPDs will also give a clue. If his sire and dam have low birthweight EPDs, he'll probably sire calves that are smaller than average at birth. You want calves that are easily born, especially if you have no experience delivering calves. The biggest factor in determining whether a calf has difficulty being born is the size of the calf at birth.

Heifers should always be bred to a dependable low-birthweight bull that sires calves that are small at birth. Birth size is partly determined by nutrition of the cow during pregnancy and whether it is her first calf (older cows tend to have bigger calves than first-calf heifers), but it is mainly determined by genetics. To play it safe, do not use a bull whose calves are heavy at birth.

A heifer having her first calf is not as big as a mature cow. She still has some growing to do. If the calf is too large, it may die during the birth process or injure the heifer. A big calf usually needs help being born. It may have to be delivered by Caesarean section, in which case the veterinarian cuts through the cow's abdomen and into the uterus to take out the calf and then stitches the cow up again. This is not the way you want the calf to be born!

Artificial Insemination

If you can tell when the female is in heat, you can have her bred AI. The AI technician can order semen from a bull of your chosen breed. There are several breeding services that collect semen from outstanding bulls across the country. Many purebred breeders, ranchers, and dairymen use AI. Most dairymen use AI exclusively and do not buy bulls; they prefer to use the most outstanding bulls in the nation. AI can be a way to use some of the very best bulls, whether beef or dairy, without having to keep a bull. Dairy bulls are very aggressive and dangerous, so it's better if you don't even need one of these.

Artificial insemination is the process of placing semen into a cow's uterus at the proper time to cause her to become pregnant. A large number of cows can be bred to one bull using AI — many more than he could sire by natural service. The semen is collected and divided into several small portions, put into small tubes, which are often

A COW'S REPRODUCTIVE SYSTEM

The cow's reproductive tract consists of the vulva (external opening), vagina (tube that connects the vulva with the uterus), cervix, uterus, ovary (where the eggs grow), and oviduct (tube between the ovary and the uterus). The cervix is the opening into the uterus. It stays shut most of the time and protects the uterus from infection. It seals off tightly while the cow is pregnant and opens at the end of pregnancy (start of labor) so the calf can come out. It also opens during heat when the cow is ready to mate with a bull.

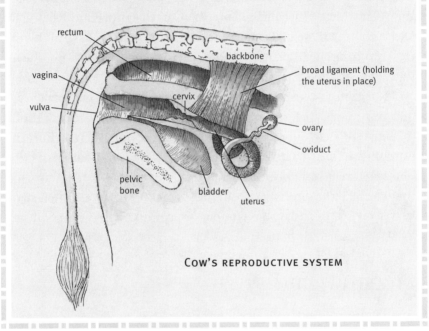

COW'S REPRODUCTIVE SYSTEM

called straws. These are stored in liquid nitrogen, which keeps them very cold (–320°F). The frozen straws of semen can be shipped anywhere.

Price of the semen varies. Some bulls, especially the most popular ones in their breed, are expensive. You don't need the most expensive semen; any good bull with genetic qualities (and low birthweight traits) you want for your calves will be fine.

Watch cows or heifers closely during breeding season (a few weeks in June, for instance, if you want calves born in March) to detect when each one comes into heat. A cow or heifer will probably be in heat for about 12 to 18 hours, though some are in heat for a shorter or longer time. Try to spend at least 30 minutes twice a day, morning and evening, watching the cattle for signs of heat. When you see an animal in heat, you can call the AI technician.

With the cow restrained in a chute, semen is inserted into her uterus through the vagina and open cervix. The technician uses a long slim tube to deposit semen in the proper place. With good luck the cow will settle (become pregnant). If she does not conceive, she'll return to heat 18 to 23 days later and must be bred again. Watch her closely during the time she might return to heat. If she does not show signs of heat, she's probably pregnant.

For a group of cows or heifers, you might consider heat synchronization. Many stockmen who utilize AI for breeding use this method to breed the herd about the same time. This eliminates the chore of watching them all so closely during several weeks of breeding season. The cows are given hormone drugs that "time" their heat cycles, so they can all be inseminated about the same time. If you are interested in this procedure, ask your veterinarian, county Extension agent, or local AI technician about it.

Prenatal Care

Good pasture gives cattle all the nutrition they need for pregnancy. You won't need to increase their ration right away if you are feeding hay instead of pasture; the fetus in early pregnancy does not require much. The last stage of gestation (third trimester) is when a cow's nutritional needs increase. Excessive overfeeding then may cause the fetus to be larger than normal, but this is generally not a problem. If pregnant cows are nursing this year's calf, they need more feed. Lactation (producing milk) requires a higher level of nutrition than does pregnancy. Make sure cows stay in good body condition and heifers continue to grow, without becoming too thin or too fat.

As the fetus grows, the pregnant cow's nutritional needs increase. At two months' gestation, the fetus is the size of a mouse. By five months, it is the size of a large cat. It then grows rapidly during the last two months, becoming a full-size calf by nine months. Adjust feed as needed.

If your winter climate necessitates feeding hay, keep close watch on pasture through fall and start feeding hay as soon as the best pasture is gone or snow-covered. Pregnant dry cows (current calves weaned off and the cows are no longer producing milk) usually do well on good grass hay. If grass hay is poor or cows are thin, however, a mix of grass and alfalfa will be better for them. Heifers pregnant for the first time are still growing and do better on a mix of grass and good alfalfa hay. Start feeding them alfalfa hay when pasture gets short and increase the hay as needed. Alfalfa/grass hay is usually the only type of feed they'll need through winter.

Pregnant heifers that are gaining weight and growing won't need grain. Most beef heifers do not need grain during pregnancy, unless they start to get thin. This may happen if weather is severely cold for a long time, or if the hay is not good quality. Feed all the good hay that heifers will clean up. If they are eating every bit of it, increase the ration. If they start to waste very much, cut them down to the amount they will readily eat.

Cattle that are extremely eager at feeding time — bawling and waiting at the gate — are hungry and not getting quite enough to eat. Cows that are content and more nonchalant about coming to the

WEANING CALVES FROM PREGNANT COWS

Calves are generally weaned before the cow enters her last trimester (six months pregnant — when her suckling calf is eight to nine months old). This years' calf should always be weaned at least two to three months before the cow calves again.

feed are getting the right amount. If they are wasting hay and bedding on it, you are feeding too much. It's cheaper to cut back on the amount of hay and give them straw to bed on instead.

Good grass hay provides roughage and nutrients a cow needs, while good alfalfa hay can provide the extra protein, calcium, and vitamin A necessary for growth and pregnancy. If the hay you are feeding is deficient in protein or vitamin A, provide a supplement. Your county Extension agent can send a sample of your hay to a laboratory for testing to see if it is adequate or whether you need a supplement.

Cattle need more feed in cold weather. Increase the amount of hay, especially the grass hay. During cold weather it takes more calories to provide body warmth, and grass hay is best for this because it creates more heat during digestion.

During the last three months of gestation, a pregnant animal needs more feed. Feed cows and heifers well, but don't let them get fat. If cattle are on hay during this time, increase the hay. If calving season is in early summer, good pasture can provide the extra nutrition needed in the final stage of pregnancy.

Cows on pasture do better than those confined in a small pen; they exercise more by walking around and grazing. An overly fat cow or heifer that doesn't exercise during pregnancy may have a harder time calving; she will tire during labor, and globs of fat around the birth canal can make it difficult for the calf to come through.

Vaccinations

Cows and heifers should be on a vaccination schedule to receive booster shots for certain diseases once or twice a year. Some vaccines can be given during pregnancy and some should not. Talk to your veterinarian about vaccinations your cattle will need.

Ask your veterinarian about vaccination to help protect young calves against infectious types of scours. If your pens or pastures have ever held baby calves and have been contaminated with diarrhea, or you have neighbors that have had problems with calfhood diarrhea, it's wise to vaccinate cows and heifers against types of disease that are preventable. This enables the pregnant animal to create antibodies against those disease organisms. The cow can pass antibodies to her new calf when he nurses; her first milk (colostrum) will contain high levels of protective antibodies.

Get Ready For Calving

As calving time approaches, make sure you have a safe place for cows or heifers to give birth. You don't want them to slip on ice or a muddy hillside or get stuck in a ditch.

The calving area should be clean. A green pasture with no gullies or ditches is often the safest and cleanest place for a cow to calve. In cold, windy, or stormy weather, provide a pen or shed where there's more shelter; make sure there is plenty of clean bedding and change the bedding when it gets soiled. You don't want her calving in mud or manure, or even on dry bare ground in a barnyard where there have been lots of cattle. The calf must be born in a clean place or he may get an infection.

CALVING KIT

Things to have on hand at calving time:

- Halter and rope, in case you need to tie a cow or heifer
- Strong iodine (7 percent solution) in a small wide-mouth jar, for dipping a calf's navel (see page 137)
- Towels, in case you have to dry a calf
- Bottle and lamb nipple, in case you need to feed the calf
- Obstetrical chains or short, small-diameter (half-inch) smooth nylon rope with a loop at each end, in case you have to pull a calf
- Disposable OB gloves (from your veterinarian)
- OB lubricant in a squeeze bottle
- Flashlight for checking on cows at night

Navel ill, sometimes called joint ill, can be caused by bacteria entering the calf's navel stump soon after birth. If cows are not calving on grassy pasture, have some straw on hand so you can supply clean bedding and keep the calving area very clean.

Watch for signs that she will soon calve. As a cow or heifer approaches her calving time, her udder will fill. The vulva (where the calf will come out) becomes large and flabby. Those muscles are relaxing so they can stretch wider as the calf emerges. The area between her tail head and pinbones becomes loose and sunken. These changes may start several weeks or just a few days before the cow calves. Some cows or heifers have a large udder for a long time. Others may "bag up" almost overnight.

Stages of the Birth Process

The calving process has several stages. Knowing what occurs at each stage will help you decide when or if a cow or heifer needs help. Most cows have no trouble unless the calf is in the wrong position to be born, but heifers might have a hard time if they are trying to give birth to a large calf.

Early Labor

When the fetus has finished growing and is ready to emerge, hormonal changes signal the cow's body to go into labor. Signs of first-stage labor are restlessness and mild discomfort. The cow has a few early contractions as the uterus prepares to expel the calf. She may kick at her belly or switch her tail.

Contractions become more frequent and more intense as labor progresses. Contractions of early labor help turn the calf toward the birth canal. The calf is usually positioned with front legs aimed toward the birth canal and his head resting between them. If he is in a different position, the contractions of early labor usually help him move into the proper position.

First-stage labor may last two or three hours in a cow, but might take four to six hours or longer for a heifer having her first calf. She

Amnion sac emerging; water sac (which preceded it) is hanging lower.

may pace the fence, trying to get out of the pen or pasture. She may go into the bushes or a secluded corner. It's her natural instinct to go where other cows won't bother her or try to steal her calf.

Second-Stage Labor

When the cervix is fully open and the calf or the water sac — often pushed ahead of the calf — enters the birth canal, active (second stage) labor has begun. The birth should take place within one-half to two hours.

The water sac is dark colored and purplish. When it breaks, dark yellow fluid rushes out. The sac may break before it comes out and all that you'll see is fluid pouring from the vulva. The water sac usually comes ahead of the calf, but it sometimes comes out intact right after the calf. It should not be confused with the amnion sac enclosing the calf. The amnion is a thin, white membrane full of thicker, clearer fluid. These sacs surrounded the calf while he was inside the uterus, buffering him from bumping or injury.

Active labor is more intense than early labor. The cow has strong abdominal contractions. Entrance of the calf into the birth canal triggers hard straining. Each contraction forces him farther along. Soon the feet appear at the vulva. It may take awhile for the vulva to stretch enough to pass the calf's head. The calf can safely be in this position for one to two hours, but it is best if he can be born within an hour or less. Give a first-calf heifer time to stretch her tissues, however. Pulling forcefully on the calf too soon may injure her.

The cow or heifer may be up and down a lot, but once she starts straining hard she will generally stay down. If she is making progress and the calf is gradually coming, just stay back and quietly observe. If she is in a pen or barn, make sure she doesn't lie up against the fence or stall wall; you want to be sure there is room behind her for the calf to emerge.

It may take awhile for her to pass the calf's head, but as long as she is making progress, you don't need to help. After the head emerges, the rest of the calf usually comes easily. Fluid flows briefly from his mouth and nostrils as his shoulders pass through the vulva and his ribcage is squeezed through the cow's pelvis. This fluid was in his stomach and air passages while he floated around in the uterus, but now comes out so he can start to breathe.

After the calf is born, the cow may lie there a few moments to rest. A cow that's had calves before will usually jump up, turn around, and start licking the calf, but a first calf heifer is usually tired and may be slower to realize she's had a calf.

The calf must begin breathing immediately, however. Usually the amnion sac that was over his head will tear during the birth process; the fluid will drain away and he can start to breathe. Sometimes it doesn't though, and the calf may suffocate. A cow that jumps up and starts licking the calf may lick the sac off in time, if she starts licking on the proper end of the calf. The calf may die if she does not. If you are watching and see the sac is still over the calf's head or still full of fluid, you must quickly help him. Pull the membrane away from his nose. Clear the fluid away and make sure he can start breathing.

Cow licking newborn calf.

Shedding the Afterbirth

The third and final stage of labor is when the cow sheds the placenta, often called the afterbirth. When she gets up, there should be a lot of red tissue hanging down from her vulva. This is the placenta, which surrounded the calf in the uterus and attached to the uterus walls. The attachments ("buttons") are dark red, dollar-size objects spaced over it.

It may take 30 minutes or even a few hours for the placenta to completely detach from the uterus and come out. When it does come out, you should remove it from the barn stall or pen so the cow won't choke on it. Cows usually eat the placenta so it won't attract predators. If the cow calves out on pasture and you are not there to observe the birth, you probably won't find the placenta because she will already have eaten it.

If it takes longer than 10 or 12 hours for a cow to shed the placenta, she may develop a uterine infection. Keep close watch for signs of fever or illness. She may develop a fever, becoming dull and breathing fast, or refuse to eat and need immediate treatment. Call your veterinarian. Never pull on the placenta while it is still hanging from the cow. It must come away by itself. If you pull on it, it may break off, leaving part of it still inside the uterus to cause an infection. Some cows that retain the placenta do not get sick, and some cows that shed the placenta normally will still develop an infection. Watch the cow closely for several days after calving.

Helping a Cow or Heifer Calve

Sometimes you must help during the birth process. The calf may be in the wrong position in the uterus and so can't enter the birth canal or come through it. Maybe he's a little too big to come through easily, or maybe the cow is having twins. If labor continues for too long and nothing is happening, she should be checked. Call your veterinarian or an experienced person to help. Definitely call for help if you see *only one foot* coming out, or if you see hind feet instead of front feet.

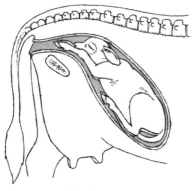
Normal birth presentation

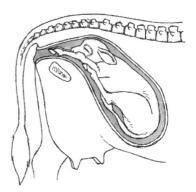
All four feet coming into birth canal

Checking Inside the Cow

When there's a problem with the birth, a careful examination inside the cow may be necessary. If she is gentle, you can check her yourself. First tie her up or restrain her in a head catcher or stanchion so she can't move around. If she is lying down and does not get up when you approach her, you can probably check her right where she is. Take care to be as clean as possible, to avoid introducing infection. If her vulva has manure on it, rinse it off with clean, warm water. Use a disposable long-sleeved plastic glove that covers your whole arm, if you have one. Otherwise, just wash your hands and arm and apply OB lubricant to your hand and forearm.

If nothing has appeared at the vulva yet, feel into the birth canal to see if you can find two feet. If the feet are there, check their position to see if they are front feet or hind feet (see page 135). If there is just one foot, the nose of the calf and no feet, or some other abnormality, you'll know why the birth is not progressing.

If nothing has come into the birth canal yet, feel farther and examine the cervix with your fingers. If it is not open yet (you come to a closed end of the birth canal), or is only partially open (you can only get one or two fingers through), you are interfering too soon. When the cervix is completely open, it will be six to seven inches wide and you can reach clear into the uterus.

If the cervix is open, and your hand can go clear through it, reach into the uterus and check the calf. You'll probably feel his feet. If he is not positioned properly, the first thing you touch may be his head, tail, some other part of his body, or one foot. You need help immediately to reposition the calf so he can be born.

Pulling a Calf

Often the only problem is that the calf is too big and needs a pull. Do not pull on a calf unless he is in proper position to come. If the feet have been showing for an hour and you've felt inside the vulva and the nose is right there, you know the head is coming out properly and is not turned back. Go ahead and pull the calf.

If feet are sticking out, the calf's nose is showing, and the heifer's straining starts to push the head out, you can wait. She probably won't need help. If she is not making progress after the feet have been showing for an hour, help her. Although she might eventually have the calf, the prolonged effort will wear her out. It will take longer for her to recover from the birth and she may even be paralyzed from pressure on the nerves that go to her hind legs, and the stress and pressure on the calf will be hard on him. After a prolonged time in the birth canal he may not be able to stand or nurse. His head, mouth tissues, and tongue may be swollen from all the pressure, and this will prevent him from nursing on time. It's better to help with the birth.

First feel inside the birth canal to make sure there is room for the head to pass through the pelvic opening. If you cannot fit your fingers between the top of the calf's head and the top of the birth canal, the opening may be too small. If this is the case, call the veterinarian.

If you think the head can come through, go ahead and pull on the calf's legs. It helps if two people work as a team to pull the calf. Pull

Stretching the vulva

alternately on one leg and then the other, to ease the calf through the pelvis one shoulder at a time. This will often start him moving, since he is not so wide that way.

The calf has to come in an arc. He starts out down in the uterus and must come up and over the cow's pelvis to enter the birth canal. When his feet emerge from the vulva you should pull straight out. After his head comes out, pull slightly downward, and as his shoulders and ribcage emerge you can start pulling downward toward the cow's hocks. If you watch a normal birth, you'll see the calf curves around toward the cow's hind legs as she is lying there and he slides out.

It's difficult to pull a calf with your hands because his legs are so slippery. It's better to use obstetrical chains with handles that hook into the chains at whatever point you need to be pulling. If you plan to raise cattle, buy a set of chains from your veterinarian. If you don't have chains, put a rope or baling twine around the calf's legs and use gloves when pulling, so the rope or twine won't cut into your hands. Whether using chains or a rope, position the loops above the fetlock joints so they won't injure the joints or the feet when you pull.

In a difficult birth, one person can pull on the front legs of the calf while the other stretches the cow's vulva. This helps the head come through more easily. One person pulls while the other stands beside the cow, if she's standing, or sits beside her hips if she is lying down, facing to the rear. When stretching her vulva, put your fingers between the calf's head and the vulva, pulling and stretching the vulva each time the cow strains. You and your partner should pull and stretch the vulva only when she strains, and wait while she rests. Don't pull when she is not straining, or you may injure her.

Hiplock

Sometimes in a hard birth, you pull the calf partway out, only to have him stop at the hips. Don't panic. Remember he has to come up and over the pelvic bones in an arc. As his body comes out, start pulling downward, toward the cow's hind legs. To avoid hurting his ribs, move him out far enough that his ribcage is free before you start pulling downward. Once his ribcage is out, he can begin to breathe if the umbilical cord is pinching off (which it may do because of so little room inside the birth canal).

If the cow is standing, pull the calf straight downward and underneath her, between her hind legs. This raises his hips higher to where the pelvic opening is the widest. He can come through much more easily than if you try to pull him straight out and his hips are caught against her pelvic bones. If the cow is lying down, pull the calf between her hind legs, toward her belly.

Backward, Breech, and Other Position Problems

A calf coming backward is usually not born alive unless you help him. A backward birth is slow because the calf is not very streamlined in this position. His head is still inside the cow when his umbilical cord breaks, and he cannot breathe. He dies of suffocation.

If hind feet protrude from the vulva when you observe the cow, this is an emergency situation. You can tell they are hind feet because the bottom of the feet and the dewclaws will be facing upward. Front legs have toes pointing down, and dewclaws are on the bottom. If the bottoms of the feet are up, the calf may be backward. But before you assume this, feel inside the birth canal to tell if there are knees (front legs) or hocks (hind legs). Sometimes the calf is coming frontward but is twisted or sideways — or upside down.

If the calf is backward, call the veterinarian, or call a close neighbor who has calf pullers. You will need a pulling winch to get the calf out before he suffocates. The umbilical cord breaks as the calf's hindquarters come through the pelvis, and since the calf must then start breathing, he must come out fast. Pulling him out by hand is usually not fast enough to save him.

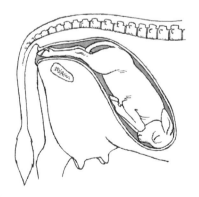

Posterior presentation

A breech calf is backward, but the legs do not enter the birth canal. He is coming rump first — too large to start through the cervix. With the calf

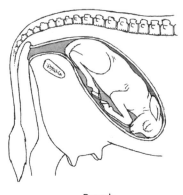

Breech

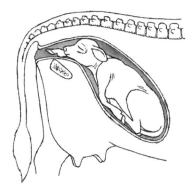

Calf with one front leg turned back

in this position, the cow may not start second-stage labor because there is nothing in the birth canal to stimulate hard straining.

The only sign of a breech presentation may be that the cow just seems to be taking a long time in early labor. If you wait too long before checking her, the placenta will detach from the uterus and the calf will die. If you check inside the cow, however, you will be able to feel the rump or tail. You will need a veterinarian or an experienced person to push the calf forward enough to get each hind leg, one at a time, into the birth canal so the calf can be pulled with a calf puller.

Sometimes a calf is properly positioned but one front leg will be turned back, so that only one foot appears at the vulva. Sometimes the head and one front foot will show. It's best if you detect this problem early, before the head is pushed out too far. That way the calf can be pushed back into the uterus where there is room to straighten the other leg so he can start coming properly. Get help immediately.

After the Calf Is Safely Born

There are several important things to do after the calf arrives. First, make sure he starts breathing. Then disinfect the navel stump, unless he is born in a clean grassy pasture where there is little risk for infection. Finally, make sure he gets up and nurses his mother within a couple hours of birth.

Tickle a newborn calf with straw to get her to start breathing.

Start Breathing

After any delivery, make sure the calf starts breathing as soon as possible. Try to stimulate breathing by sticking a clean piece of hay or straw up one nostril. This tickling will usually make him sneeze or cough and he will start breathing.

If there is no response, the calf may be unconscious from a prolonged lack of oxygen during a hard birth. Give him artificial respiration. This can keep a calf alive by putting oxygen into his lungs to revive him. If he is still alive, you can feel his heartbeat if you put your hand on the left side of his ribcage, behind his front leg. If there is no heartbeat, you are too late. As long as there's a heartbeat, there is hope for the calf, and you should try to blow air into his lungs.

To perform artificial respiration on a calf, blow a full breath of air into one nostril while holding the other nostril and his mouth closed. Blow until you see his chest rise. Then let the air come back out. Blow in another breath. Keep breathing for the calf until he regains consciousness and starts breathing on his own.

Disinfect the Navel

If the calf is born in a barnyard, pen, or barn, disinfect his navel stump as soon as the cord is broken. Have a small wide-mouth jar (like a baby food jar) ready ahead of time, with half an inch of strong (7 percent) tincture of iodine in it. Immerse the navel stump and make sure the whole stump is completely soaked in iodine by holding the jar tightly against the calf's belly for a moment.

Disinfect the navel stump with a small jar of iodine.

Be careful with iodine. Avoid spilling any on your hands or on the calf; it is a strong chemical and can burn the skin. Never get any of it in your eyes or in the calf's eyes.

Nursing Newborns

Watch to see if the new calf is able to get up and nurse. If he had a difficult birth, he may be slow. If he doesn't get up within 30 minutes, help him stand. Make sure he nurses within a couple hours of birth; the sooner the better. First-calf mothers often are nervous about new calves and may not stand still at first to let him find the udder. Usually a calf will eventually catch up with her udder, but in some cases you may have to help. If the heifer is in a pen or barn and won't stand still for the calf, feed her a little alfalfa hay. She will usually stand still and eat it, giving the calf a chance to nurse.

Check the udder. If the calf was born out in a pasture when you weren't watching, you don't know exactly when he was born. You can't tell whether he nursed, unless you examine the cow's udder closely. Sometimes you can tell; the teats may be smaller and empty. If the cow had more milk than the calf could hold at first nursing, one or two of the teats may still be full and the others empty. You will usually be able to tell if he nursed.

The cow's first milk (colostrum) is very important for the calf. It has twice the calories of regular milk and creamy fat that is easily digested and high in energy. It gives the calf strength and keeps him warm if he's born in cold weather. After he gets a tummy full, he may feel like bucking and playing. Colostrum also has a laxative quality that helps him pass his first bowel movements.

The antibodies in colostrum are vitally important to a newborn calf. Calves are born with no protection from diseases, so their antibodies must come from colostrum. This temporary immunity lasts a few weeks, until the calf's immune system starts making its own antibodies. A calf that gets no colostrum, or does not nurse until he is several hours old, may have little resistance to disease and can develop scours or pneumonia in his first weeks of life.

For a short while after birth, the calf can absorb antibodies from colostrum directly into his bloodstream through pores in the intestinal lining. The best time for this absorption is during the first two hours after birth (preferably within the first 30 minutes), before the intestinal wall begins to thicken and the pores close up. By the time he is six hours old he can only absorb a fraction of what he needs; by 12 hours old it is usually too late to absorb any appreciable amount.

Tie the first-time mother to keep her from kicking her calf.

If he can't nurse on his own, such as if he is slow to get up after a difficult birth, help him latch onto a teat, or milk a little from the cow's udder and feed it to him with a bottle.

Nursing is easier with cooperative cows. Cows that have had calves before are usually very good mothers unless they have a physical problem (like an injured teat) that causes pain when the calf tries to nurse. Most heifers are good mothers, but sometimes a first-time mama is confused or indifferent. She may have no interest in the calf, or doesn't want to mother it, or won't let him nurse. Usually after the calf has nursed once or twice, the young cow becomes more interested in the calf; the nursing stimulates hormones that make her feel more motherly.

If she kicks at him and he can't nurse, tie her up or restrain her in a stanchion, and tie a hind leg back so she can't kick. Leave enough slack in the rope that she can stand comfortably on that leg, but not enough that she could kick. Then you can help the calf without her kicking him or you. He may go right to the udder, but if he is weak from a hard birth or timid because he's been kicked, guide him to the teat and put it into his mouth. Once he gets a taste of milk he will usually suck eagerly, and the cow may relax and cooperate.

You may need to put hobbles on the back legs of a cow to keep her from kicking her calf.

Sometimes a heifer keeps kicking. She may have a lot of swelling in her udder (called "cake"). Nursing is painful. You may have to put hobbles on her hind legs for a few days so the calf can nurse without being kicked.

If the calf has trouble nursing or is too weak to stand and nurse, milk colostrum from his mother to feed him. You need about a quart. Pour it into a small-necked bottle and use a lamb nipple for feeding it to the calf. This will usually give him the strength to get up and nurse. If he is too weak to suck a bottle for his first nursing, have your veterinarian or an experienced person show you how to use an esophageal feeder to put the milk through a tube down his throat and into his stomach.

Some cows are very protective mothers and charge at anything that comes near their calf. Never allow a dog near cows that have new calves. The cows consider it a predator (even a dog they know and tolerate under ordinary conditions), and will charge after the dog. A cow's instinct is to protect her calf; she may be dangerously aggressive for a few days.

When you handle her calf to iodine its navel, have a stick at hand to rap her across the nose if she tries to butt you. After the calf is a few days old, she will not worry so much about you handling the calf because she knows and respects you, but try not to upset her with strange people or a dog.

Avoid distractions for your cow and her new calf. Don't let a lot of people around them or invite friends to see the new calf right away. Strangers nearby might make the heifer nervous and she might not mother the calf.

Once her baby is born, a first-calf mama usually takes her new job very seriously. It's quite a change to see a sassy heifer become a concerned mother, licking her baby, mooing at him, worrying about his every movement, and not letting him out of her sight. Over the next hours, her calf will become very self-assured, bouncing around for the sheer joy of being alive. There's nothing more fun than watching new baby calves!

CHAPTER TEN

Beef Herd Management

THIS CHAPTER LOOKS AT managing a group of beef cows and the tasks involved in caring for them throughout the year. These guidelines will be useful, even if you have just one or two cows.

Winter

By this time, calves from last spring are weaned and vaccinated. If you didn't sell them as weanlings, you are keeping them to sell next summer as yearlings or beef animals. Or you may be keeping heifer calves as future cows. They should grow well through winter.

Weaned calves do better on pasture than confined in a corral for winter. They exercise more and stay cleaner. In many winter climates, a pen will become muddy; it may be knee deep in mud and manure by spring. Calves get cold standing in mud to eat at a feed rack. Frozen mud creates difficult footing and can lead to foot injures and foot rot. They will be more comfortable and stay healthier out on pasture, even if there is no grass to eat. If you feed hay on the ground, choose a clean place each time. If cattle must be confined in a pen, make sure they have a dry, clean place to sleep. Put straw in the bedding areas.

Winter feed

You hope your cows and yearling heifers are pregnant at this time. Your job is to bring them through winter in good shape for spring calving. Young cows that are still growing probably need alfalfa hay along with their pasture or grass hay. Mature cows can get by on grass hay. If your climate has little snow, and there's still pasture available, mature cows just need a protein/vitamin supplement to make up for deficiencies in winter grass. Always provide salt. Give trace minerals if your feeds are short on these. Make sure water supplies are adequate and not freezing up. Break ice daily or use a water tank with a heater.

Delouse

Lice on cattle multiply more swiftly in cold weather when hair coats are long. Lice are not as much of a problem in southern areas (cattle don't often carry heavy infestations), but can be an irritating and costly parasite in northern regions.

If there are any lice on your cattle, delouse the whole herd in late fall and again in late winter. Since lice are easily spread from animal to animal, it doesn't do much good to delouse just part of the herd or let treated cattle mingle with untreated cattle. Do them all about the same time, even if they are in separate groups. Talk to your veterinarian about a good control program and appropriate products to use.

Winter diseases

Keep close watch for wintertime problems such as foot rot, or weather/stress-related illnesses such as pneumonia. Cows and heifers should already be vaccinated against leptospirosis, IBR/BVD, and redwater, if it is a problem in your area. If you are vaccinating cows to create antibodies in their colostrum to protect their newborn calves against scours, the best time to give these shots is several weeks before calving. Heifers expecting their first calves may need two shots, several weeks apart, depending on the type of vaccine that is used. Cows vaccinated the year before only need a booster shot. Talk with your veterinarian ahead of time and figure out an appropriate vaccination schedule that will fit your herd.

Spring

Spring is the preferred time for calves to be born, after the cold weather and mud of winter are past and there is enough green grass to pasture cows without having to feed hay.

If you have several cows or heifers, tag the calves as they are born. That way you'll never mix-up calves in records or other management procedures. Before calving, buy ear tags and number them. If the cows have numbers, give each calf its mother's number. This makes your record keeping and cattle management much easier. Most stockmen then retag their replacement heifers at weaning time to give them their own permanent number, since some use a numbering system that indicates their year of birth. Other stockmen use brisket tags for the cows, which are much more permanent than ear tags, and ear tags (same number) for the calves.

Purchase any medications and supplies you might need, and have them on hand for calving (iodine for navels, vaccine for calves, obstetrical gloves, scour medications, and so on). Then if a cow or heifer calves a couple weeks early, it won't catch you by surprise; you'll be prepared.

Cow/Calf Management

If cows calve in a large, clean pasture, they usually don't need much management except to make sure calves are safely born, tagged, and vaccinated at birth. The cow or heifer usually goes off by herself to calve and has no distractions from the other cattle to interfere with mothering and bonding with her calf. If cows calve in a confined area, however, it is wise to have a separate calving pen or two in which to put a cow or heifer by herself when she calves. This way another cow won't steal the new baby, and the newborn calf will learn to identify his mother. He won't become confused and try to nurse the wrong cow and get kicked. Once the pair has bonded, they can be turned out with the herd into a different pen or pasture from the pregnant cows.

Keep cows with babies separate from the cows that have not yet calved. They need to be fed differently; cows need more feed when pro-

Young calves need a shed to protect them from bad weather.

ducing milk than they do when pregnant. You also want to keep the calving area very clean, with no calves in it. If young calves become sick with diarrhea, the pen or calving pasture quickly becomes contaminated. A pregnant cow may lie where sick calves have been and get her udder dirty; she then could pass the harmful pathogens to her newborn calf when he nurses for the first time.

Make sure every newborn calf has colostrum within two hours of birth. Freeze extra colostrum for use in emergencies. If a gentle older cow has a lot of colostrum when she calves, steal a quart from her while her own calf is nursing. If you squat down alongside her own calf, you can reach to the other side of her udder and milk into a jar or pitcher. A small plastic pitcher is easy to hang onto with one hand while you milk with the other.

Freeze the colostrum in a plastic container and save it for later. If you ever have a calf that can't nurse its mother for some reason, or a cow that does not have enough colostrum, you'll have some. Colostrum will keep for several days in the refrigerator and for several years in the freezer.

Don't thaw colostrum in a microwave or get it too hot. Excessive heat destroys antibodies and vitamins. Put the container in hot water to let it thaw, and never heat colostrum much higher than calf body temperature. When you feed it to the calf, it should feel comfortably warm, but not hot, on your skin.

Young calves need protection from bad weather. If they are cold and wet, they are susceptible to scours and pneumonia. Build a small shed or portable calf house that calves can enter, but cows can't. Make a low entrance (so only calves will fit) or put panels around the front with poles the calves can go under. Then cows won't try to go in the shed or lie in front of it, and possibly lie on a calf.

Navel Care

Navel ill is a serious infection that can kill or cripple a calf. Bacteria that enter the navel stump after birth may create an abscess in the navel area or get into the bloodstream and cause general infection (septicemia), which can be fatal. Bacteria may settle in the joints to cause a septic arthritis, in which case the calf has painful joints and is very lame. It can be difficult to save a calf once it gets navel ill, even with diligent treatment.

Prevent this devastating disease by making sure calves are born in a clean place and by disinfecting the navel stump of each calf immediately after birth with strong iodine. The iodine not only kills germs but also acts as an astringent, shrinking the tissues and helping the navel stump dry up and seal off quickly so bacteria cannot enter.

Don't touch the navel cord with your hands unless they are very clean. The only time you need to touch the cord is if it is too long and drags along the ground when the calf is standing. Then you should cut it with very clean, sharp scissors, leaving about a three-inch stump. Be careful to not jerk on it. If you pull on the cord too much, this could injure the calf internally. As soon as you cut it, immediately soak the navel stump in iodine.

The navel stump of a baby heifer usually dries up after just one application of iodine. A bull calf's stump may take longer. He urinates very close to the navel, and if he urinates while lying down, as many baby bulls do, he keeps wetting the navel stump and it won't dry up as fast as it should. As long as it is still wet, bacteria can enter it easily, so reapply iodine a few hours later, and again, if necessary, until the stump is dry. You want it to dry up completely by the time the calf is 24 hours old.

Occasionally a calf will have a navel (umbilical) hernia, in which case the abdominal wall at the navel area will not close up properly after the calf is born. There will be a bulge at the navel. This should not be confused with an abscess. Navel infection sometimes causes an enlargement, but an abscess will be very firm. A hernia is soft tissue and may go back up through the hole in the abdominal wall if you press it. If a calf has swelling at the navel, have your veterinarian check it. An abscess should be lanced, drained, and flushed with antibiotics. If it's a hernia the veterinarian can tell you whether it is harmless or needs treatment.

A small hernia may go away by itself as the calf grows; the tissue will no longer come through the small hole. A large hernia can be serious. A loop of intestine may come through it and strangulate. The piece of intestine gets pinched; the blood supply to it is hindered. This may cause the segment of intestine to die, killing the calf. If you discover it early, however, your veterinarian can stitch the hole closed before it becomes a problem.

Newborn Medications

Your veterinarian may recommend certain medications for newborn calves, depending on your region and situation. If feeds were short on vitamin A through winter and cows did not have a supplement, the veterinarian may recommend a shot of vitamin A for each calf unless it's late spring and green grass is growing well.

In some areas, newborn calves need an injection of selenium to avoid "white muscle" disease. You may also need to vaccinate young calves against clostridial diseases such as enterotoxemia (a highly fatal gut infection) or tetanus, if these diseases are a problem in your area or herd. Discuss calfhood medications and treatments ahead of time with your veterinarian so you will be prepared and have the proper medications or vaccines on hand when calves arrive.

Scours

Infectious diarrhea kills more young calves than any other disease. Bacteria and viruses that cause scours may be lying on the ground, just waiting for the right conditions such as wet, muddy weather. Some types of scour bacteria can last a long time, lurking in old manure or bedding from sick calves in earlier years. Sometimes bacteria or viruses are brought into a pen or pasture with purchased calves, or carried to your farm from a neighboring farm on the feet of birds, animals, or people. A young calf can easily pick up harmful pathogens when nibbling dirt or mud, drinking from a puddle, licking himself after lying in mud or other calves' manure, or nursing a dirty udder.

The key to preventing scours is diligent good management: maintaining a healthy cow herd, uncontaminated areas for calving, and clean bedding so cows don't get their udders dirty, and making sure every calf gets an adequate amount of colostrum soon after it is born. Pre-vaccinating cows ahead of calving can help prevent some types of scours but not others. Antibodies from vaccination may not be enough protection if cattle are in a dirty environment.

Wash the teats of any cow or heifer you assist at calving, clean all your equipment between each use (especially bottles, nipples, and your esophageal tube feeder), and move cows with new calves to a clean pasture after they have bonded.

Never put sick calves in the same barn or pen where you'll have cows calving. A calf with scours should be isolated from the rest of the group; put him and his mama in a separate pen, or in a different shed if he needs shelter. Take him out of the herd so he won't spread germs all over the pen or pasture every time he squirts feces. Scours can quickly go through a group of calves; isolating sick ones for treatment until they recover helps cut down on the number of cases.

Calf scours can be caused by bacteria, viruses, protozoan parasites, poor nutrition, or overfeeding on milk. You may need help from your veterinarian to determine the cause and how best to treat the diarrhea. Most types of infectious scours are deadliest if a calf is very young when he gets sick. He has less body mass than an older calf and is more at risk for the harmful effects of dehydration (loss of body fluid from diarrhea). His gut lining is also more fragile, and not as able to repair itself, so it takes him longer to recover. With many types of scours, a one-month-old calf may recover quickly, while the same infection in a week-old calf might kill him unless you give him intensive treatment.

When using antibiotics to treat scours, it's better to give an oral liquid than injections or pills. Injected antibiotics don't help scours very much; the medication does not get into the gut. You need the antibiotic to go directly into the digestive tract rather than the muscle and bloodstream. Pills do not dissolve quickly enough in the stomach. The calf's normal digestion is disrupted by the infection, so the pills are not dissolved very well. A liquid antibiotic is much better.

Get an oral antibiotic from your veterinarian. If it is only available in pill form, crush those and mix with a little water. Squirt the fluid into the calf's mouth with a needleless syringe (see chapter 3) or add it to the fluid you are giving by stomach tube or esophageal feeder.

The most important treatment for diarrhea is to replace fluids and body salts the calf is losing, so he won't get so dehydrated. The sick calf needs lots of fluid. Give him a quart of warm water by esophageal feeder every six to eight hours until he starts to recover. Have an experienced person show you how to put the feeding tube down the calf's throat. Every time you give him warm water, add some electrolyte salts to it. If you're giving him an oral antibiotic, it only needs to be given once a day.

Don't overdo antibiotics. If you must give it more than two or three days, it may kill off good "gut bugs" as well as bad ones. After the calf recovers, you might have to give him a bolus or paste containing the proper rumen bacteria to restore his digestive function. These products can be obtained from your veterinarian.

You can buy packets of electrolyte mix, but these are usually expensive because they also contain nutrients for the sick calf. You usually don't need this if you catch illness early, before the calf is really weak. A simple homemade electrolyte mix is cheaper, and quite adequate if a calf has not been sick very long, but the commercial packets are very beneficial if the calf is weak.

HOMEMADE ELECTROLYTE MIX

To make one dose of your own electrolyte mix for a scouring calf, mix ½ teaspoon of regular table salt (sodium chloride) and ¼ teaspoon of Lite Salt (sodium chloride and potassium chloride). If the calf is very sick, add ½ teaspoon of baking soda (sodium bicarbonate) to reverse or prevent acidosis, a condition in which the body chemistry is out of balance due to dehydration. If a calf is weak, you can add a table-spoonful of powdered sugar or honey to give him energy.

Dissolve the mixture in warm water (one quart for a small calf, up to two quarts for a large one). Add a liquid antibiotic (such as neomycin sulfate solution, sometimes called Biosul) to one of the fluid feedings each day, using proper dosage for the size of the calf, unless you are using a different oral preparation from your vet. You will generally have to give the fluid via a tube feeder since the calf will usually not want to nurse a bottle or will refuse the taste of it.

You can also add two to four ounces of Kaopectate (¼ to ½ cup) to this fluid mix if the calf has watery diarrhea. The Kaopectate helps sooth the gut and slow down the diarrhea. If you don't have Kaopectate (or a Kaolin-pectin mixture from your vet) you can use a human adult dose of Pepto Bismol.

Calves that die from most types of scours usually die from dehydration rather than from infection. They lose too much body fluid. Your main job in treating a sick calf is to prevent this. Usually, if you can give fluid and electrolytes several times during the first 24 hours — getting up in the night for one of the treatments — a calf will recover quickly, especially if you catch the problem early. In early stages of scours a calf may still be strong, but as loss of body fluids worsens, he becomes weak and dull. His eyes seem sunken into his head and his legs are cold. If dehydration is not halted and body fluids replaced, he becomes too weak to stand and body temperature drops to subnormal. His gut shuts down; he cannot absorb oral fluids;

he needs IV fluids immediately (given by your veterinarian) or he will die. Start treating calves at first hint of scours, before you must resort to drastic measures to save them. Learn to use an esophageal feeder. An esophageal feeding tube is a handy way to get fluids into a sick calf with scours or pneumonia. It is also a way to get colostrum into a newborn calf that cannot nurse. The esophageal feeder is a container attached to a tube or stainless steel probe that goes down the calf's throat and down the esophagus. Get one from your veterinarian; have him show you how to use it.

In order to treat calves at the first hint of sickness, you must watch them closely. Of all your cattle, the baby calves need the most constant observation and care because they are most vulnerable to problems. Their immune systems are not yet ready to give much protection from common disease germs, and the temporary immunity obtained from the antibodies in mothers' colostrum may not be enough; it will start to wear off within a few weeks.

If you can catch scours early enough, when calves are still strong and lively (and hard to catch!), you can often halt the infection before the calf needs fluids and electrolytes and before the gut is seriously damaged. Neomycin sulfate solution will halt many types of bacterial scours, but won't affect viral scours. You can put this liquid antibiotic into a syringe and squirt it into the back of the calf's mouth. Use 1 cc per 40 pounds, which means about 2 to 3 cc for a young calf, depending on his size.

Many clues help you spot trouble before a calf is critically ill. A calf may be dull even before he shows diarrhea. Watch for calves that quit nursing or one that lies down by himself when the rest of the babies are running and playing. If you're feeding hay, feeding time is a good time to check calves and observe them for signs of illness. Don't forget to check cows' udders. The first sign of trouble is often a calf not nursing because he doesn't feel well. If any cow has a full udder, or is only partly nursed out, or is bawling, take a close look at her calf.

Sometimes a calf is so sick he is not hard to catch. Trying to treat calves *before* they get weak can be challenging. A gentle one can be herded into a fence corner, unless it's a wire fence he can push through. You might have to bring him and his mama into a corral, or you can do a two-person sneak where one person walks in front of the calf to distract him with funny noises, arm movements, or singing. While the calf is busy watching the "clown" the other person sneaks up behind him and grabs a hind leg. Then the two of you can hold onto him and give him the medication.

Vaccinations

Cows should be vaccinated during the two or three months after calving and before rebreeding. This is the time to give shots for leptospirosis, IBR/BVD, and other diseases. If you use any modified live virus vaccines, do it at least three weeks before breeding time to give cows time to develop adequate immunity before they become pregnant, since some of these diseases cause abortions. Some live-virus vaccines also can cause abortion in pregnant cows. Give these shots during the time of year the cows are *not* pregnant, to be safe.

Young calves also need vaccinations at this time against blackleg, malignant edema, and other clostridial diseases. You can use a combination vaccine that includes the important ones in your area. Your veterinarian can recommend a good vaccination program.

Castration

All bull calves should be castrated unless they will be used for breeding. If you are raising purebred cattle you may want to select your best bull calves to keep for breeding — for your own herd or to sell to another stockman. All other male calves should be castrated, the younger the better.

Castration is harder on the calf the older he gets. As the testicles grow, blood vessels supplying them also become larger. There is always danger of bleeding when castration is done surgically, so this should be done by an experienced person.

Baby calves are easy to castrate, however, by using elastrator rings (strong rubber rings the size of Cheerio cereal). These are stretched with a special tool and placed over the scrotum of the calf, situated at the top of the sac above the testicles. Then the tool is removed, leaving the ring to constrict tightly, cutting off blood circulation and feeling below it. Testicle tissues die and the scrotal sac shrivels up and drops off a few weeks later, leaving a small raw spot that soon heals. This is often the easiest and safest way to castrate baby calves, for there is no bleeding.

To put the ring on a calf, he should be lying on his side, with someone holding his head and front leg so he cannot wiggle around, while another person puts on the rubber ring. A few ranchers "band" their bull calves at a later age (such as at weaning time), putting strong elastic tubing above the scrotum. This is somewhat safer for the calf than surgical castration at that age because there is no risk of hemorrhage. There is, however, risk for tetanus unless the cows have been vaccinated. Later castration also takes advantage of natural growth hormones during the bull's pre-weaning period. The big disadvantage to later castration, however, is the management nuisance. Young bulls in the herd are not as placid as steers and spend a lot of time and energy chasing around and mounting cows. They sometimes get hurt in the process and often "run off" any extra weight they may have gained. There is also some risk of young bulls impregnating heifer calves if they reach puberty before weaning — as some heifers and bulls will do. Most farmers and ranchers prefer to castrate bull calves at a much younger age.

Dehorning

Some breeds have no horns and you won't need to worry about dehorning. If calves have horns, however, they should be removed when calves are young and horns are small, in order to avoid stress and risk for the calf. Older calves or mature cattle with large horns may bleed excessively when horns are removed or develop infection in the horn sinus cavities, which can also be serious.

There are several methods for dehorning. A caustic paste can be used for young calves up to a few days of age. Another method that works well on baby calves is a battery-operated dehorner that cuts a small circle around the tiny horn bud with very high heat, cutting through the skin and down to the skull, severing the blood vessels and nerves. This is probably the least stressful and most humane method; once the nerves are severed, there is no pain and no bleeding. The horn bud dies from lack of blood.

Some stockmen scoop out the horn and all horn-growing tissue with a sharp tool. This is painful and also results in some bleeding. Electric dehorners are another method, often used on calves up to three or four months of age, when horns are larger than button stage. The hot dehorning iron (in the shape of a ring that encircles the horn) is held firmly against the head, over the horn, long enough to kill the horn-growing cells at the base of the horn. The heat sears the blood vessels and there is no bleeding.

It's best if dehorning and castration can be done before fly season. Flies lay eggs in wounds made by surgical castration or head wounds created when dehorning older animals. If you use rubber rings for castration or dehorn with a battery dehorner when calves are newborn, there is generally no problem.

Summer

Your chores get a little easier in summer if cows and heifers can all be on pasture. You are no longer feeding hay and calving season is over. You still need to watch cattle closely to tell if any get sick or develop problems like footrot or pinkeye. Summer is also a time for diligent fly control. If horn flies are a problem, you may want to use insecticide ear tags (see chapter 3).

Breed Cows

If you are breeding cows by AI, start watching for signs of heat. Check twice a day, during early morning and late evening. For best results, have the technician inseminate each in-heat cow about 12 hours after heat was observed. It is standard practice to breed a cow in the evening if she was noticed that morning, or breed her early the next morning if she was seen coming into heat in the evening. Keep records of dates each cow has heat cycles, when she was bred, and her expected calving date.

If a bull is used for breeding, he should be turned in with the cows nine months ahead of the date you want the cows to start calving: May 1 to start calving in early February, or the end of May if you want them to start calving in March. If you have a group of heifers to breed, keep them separate from cows and breed them to a different bull that sires smaller calves at birth. You can use the same bull for cows and heifers, however, if you know his calves are small at birth but grow fast.

If you only have a few cows or heifers and don't have corrals to keep a bull year-round, borrow or lease a bull. Make sure he comes from a reliable stockman and is free of diseases that might spread to your cows during breeding. Know his genetic background to make sure he doesn't sire huge calves if you are using him to breed heifers.

After the cows are bred, remove the bull from the herd and keep him in a separate pen or return a borrowed bull to his owner. Bulls can be obnoxious, rubbing on fences, trying to get out to fight other bulls or find more cows to breed. They are very hard on fences; you will be constantly repairing their damage. If a bull gets in the habit of smashing down fences, he's hard to keep home. It's better to have a stout corral for him when he's not with the cows. Some bulls become aggressive, and it's not safe to have them out with the cows all the time; you may be at risk whenever you are walking among the cows.

Forty-five days is long enough time to leave a bull with the cows if they all calved about the same time. If your cows are well fed and fertile, they should all be settled by that time.

Register Calves

If you raise purebreds, summer is the time to get your calves registered, if you haven't already. This should be done before they are six months old. In many breeds, each calf will need its registration number tattooed in its ear. Find out what is required in your breed for registering a calf. Send information on each calf to the breed association, along with registration fees. The older the calf, the more it may cost to register it, so get it done now.

A hay or straw stack should be protected from wet weather with tarps, positioned over the stack in such a way that moisture will run off rather than pool on top.

Fall

Fall is the time for weaning calves, vaccinations, and getting ready for winter. Actual weaning date depends on your situation — how early or late the calves were born and how much fall pasture you have. If the calves were born in March you might be weaning in October. If they weren't born until May, you might not want to wean them until November or December, depending on what kind of weather you have at that time of year.

Buy and Store Hay

The best time to buy hay is out of the field at harvest time, contracting with someone to haul it to your place for stacking. If you didn't buy hay during summer, however, you need to locate and purchase some good hay before winter.

Always look at hay before you buy it, unless you buy it from someone who will tell you exactly what quality it is. Sometimes weather conditions are less than perfect during harvest and hay gets rained on before it can be baled or stacked. Rained-on hay is never as good as properly cured hay and should be priced lower. Sometimes you can make do with rained-on hay, if it is not moldy or too dusty; it may still work for beef cattle if you feed a supplement to make up for the lower quality of the hay.

Don't buy hay that is moldy or excessively dusty or cattle may develop respiratory or digestive problems. Some types of mold can cause abortion in pregnant cows. Stack hay in a barn or on a high, dry spot where bottom bales won't get wet and cover the stack with tarps to keep top bales from spoiling due to rain or melting snow. Make the tarp "roof" slanted so moisture will run off instead of pooling and leaking into the stack. Use bales to make a ridge down the center of the stack, so the tarp is high in the middle and sloping off both sides.

Fall Vaccinations

Vaccinate all heifer calves for brucellosis (Bang's disease) if you have not already done so, if your state requires it. Cows should be revaccinated against leptospirosis and given any other vaccinations your veterinarian recommends. All calves should be revaccinated with a booster shot for blackleg, malignant edema, and any other clostridial diseases in your area, and for viral diseases such as IBR, BVD, and PI3. Your veterinarian may also recommend treating all cattle for worms, lice, and grubs.

Selling Decisions

Decide whether to sell calves or hold them over winter to sell as yearlings. Sell steer calves in the fall if you don't have extra feed to winter them, along with any heifer calves you don't plan to keep. Decide whether you want to sell them through an auction, direct to a feedlot, or to a stockman who buys weaned calves to put on pasture or cornstalks for later sale to a feedlot.

If you're not sure how to market your calves, talk to your county Extension agent or to local farmers or ranchers for advice or for names of people buying the type of cattle you are offering. If you have purebred calves, get advice from your regional breed association representative.

Buyers who send calves to a feedlot, or to a feeder/backgrounder, may pay a slightly better price for "preconditioned" calves that are weaned and vaccinated. Calves that are over the stress of weaning will not be doubly stressed by being shipped to a new home at the same time. They are less likely to get sick. Some buyers will give you a contract ahead of the sale date, specifying the price they will pay per pound at time of delivery.

Weaning

If you precondition calves for later delivery, or wean the ones to keep over winter (to sell as yearlings or keep as future cows), choose good weather if possible, to avoid extra stress. Weaning is a traumatic experience for a calf and his mother. First-time mothers are especially frantic and may try for a week to go through fences to get back to their calves. Older cows, who have gone through this before, often resign themselves to the weaning after just two or three days.

Separating calves from their mothers and putting them in a corral by themselves without any adult cattle causes much anxiety. The calves frantically miss their mothers and the security of the herd. They mill around and pace the fence, bawling, running, and stirring up dust, which irritates their lungs and makes them vulnerable to pneumonia. Any frantic activity by one sets off a chain reaction among the others. They all start bawling and traveling again, rarely taking time to eat.

It's easier on calves if they can be weaned in a grassy pasture instead of a dusty or muddy corral. A good way to wean is with a strong net wire fence between two pastures, putting the cows on one side and the calves on the other. Calves will still bawl and pace the fence, but they can see their mothers and be near them, and after a few days they are not so worried. Another way to wean is to leave a babysitter cow or heifer that doesn't have a calf with the calves when they are separated from their mothers. The adult animal is a calming influence to help them get through this emotional time.

Make sure all fences are very strong and in good repair when you wean calves. They'll try to crawl through the fence, and so will some of the mothers. If there is any weak spot, they will go through it or over it. An electric wire along the fence also helps to prevent them from trying to crawl through.

Culling Cows

Have your veterinarian pregnancy-check cows before winter. Sell any older cow that did not settle before you have the expense of feeding her hay and then find she is not going to have a calf in the spring. If cows were well fed and the bull was fertile, there is no excuse for any cows to be "open," or not pregnant. An open cow may have a fertility problem or damage to her reproductive tract due to a difficult calving or an infection, and there is no guarantee that she will breed successfully next year.

Sell any older cows with bad teeth that are losing weight or that might become thin during winter. Sell any cows that have arthritic joints or bad eyes or bad udders, or any problem that may affect their ability to reproduce or feed a calf.

First-calf heifers that are still growing while feeding a big calf sometimes do not breed back on time. They may be open or late the next year. This is why it is important to have heifers well grown by the time of first breeding. They are less likely to skip a calf.

Take Stock

Fall is a time to look back at the year's efforts and see how calves grew, how good the heifers look, and how problems and challenges were handled and overcome. It makes you feel good if you were able to save a calf that might otherwise have died at birth, or were able to keep a sick one alive with good doctoring. It warms your heart to see him now, big and healthy and sassy. It makes you eager to see next year's babies; you are ready to face any new challenges that come along.

Part of the enjoyment and satisfaction of owning cows is in caring for them, knowing they are healthy and happy. If you enjoy cattle, the reward for your efforts is not just a paycheck when you sell the calves; it is also the pleasure of watching them grow. You enjoy getting to know them as individuals. Life with cattle is never boring. There's some hard work and a few surprises, and a whole lot of pleasure.

CHAPTER ELEVEN

Selecting Dairy Cattle

I F YOU WANT TO RAISE DAIRY CATTLE, you have several options. You might want a family milk cow or a small dairy. You can purchase a mature cow or cows that have already calved and are milking, or you can buy bred heifers that will calve soon and start milking. The cheapest way to get into the dairy business, if you are not in a hurry, is to buy baby calves and raise them yourself. You can buy heifer calves that are a day or two old and feed them on bottles. If you do not already have the milking barn and facilities you need for a small dairy, baby calves can be a nice way to work into it, giving you time to build facilities while heifers are growing.

Baby calves are cheapest to buy. Weaned heifers are more expensive. A mature cow producing milk will cost even more. The most expensive way to stock a dairy is with a bred heifer that's close to calving (springer). Bred heifers of good quality are in high demand; many dairies buy replacement cows at this age and are always looking for good ones.

Determine your goals. The most expensive high-producing cow isn't necessary if you just want a family milk cow; a gentle cow of average milk production, even if middle-aged or older, can meet your needs much less expensively than a young cow. When stocking a dairy, start out with cows already accustomed to being milked. Select for gentleness as well as milking ability. Young heifers can grow into milk cows if you train them to be easy to handle as they mature. The hardest group for a beginner to handle is bred heifers. You'll have to calve them out and train them to be milked.

Selecting a Breed

You can be successful with any of the dairy breeds, but choose one that is popular in your area if you plan to sell any heifers. You want to be sure you'll have a good market for them. See pages 235–250 for a photo gallery of popular beef and dairy breeds.

Dairy Breeds

Before you buy, decide whether you want a purebred or grade (unregistered) animal. Don't select an animal just because it is registered. Registration papers will not ensure that a calf will grow up to be the best cow. Papers won't guarantee high production or good conformation. You are better off buying a good grade animal than a poor

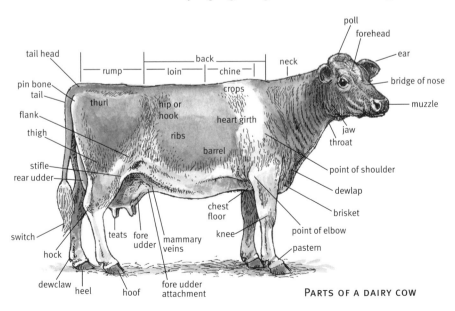

PARTS OF A DAIRY COW

registered one. When selecting the type of animals you want in your future herd, or to sell as springers, make sure they are good ones, whether grade or registered.

When choosing a purebred cow or heifer, you can use pedigree information to help you make a good choice. The pedigree is a record of that animal's ancestry. It gives genetic and performance information about her ancestors and how well the cows milked. Information about the sire and dam can help predict how a heifer will milk when she has calves. If this is the first time you've looked at dairy cow pedigrees and performance information, have the dairyman explain.

Even though pedigree information is important, common sense and good judgment are just as important when selecting a cow or heifer. Careful consideration should always be given to physical appearance of the individual animal and to her disposition and attitude. You want a gentle, smart animal that will be easy to handle.

GUERNSEY *(photo p. 249)*

Guernseys originated more than 1,000 years ago on a tiny island in the English Channel off the coast of France when the Duke of Normandy sent a group of monks to educate the islanders. The monks took along some of the best French dairy cattle, and Guernseys were developed from those. Guernseys were first imported to America in 1840. They are fawn and white, with yellow skin. The cows weigh 1,100 to 1,200 pounds (bulls weigh about 1,700 pounds), have good dispositions and very few calving problems. Their milk is yellow in color and very rich in butterfat. Heifers mature early and breed quickly.

AYRSHIRE *(photo p. 248)*

These cattle originated in Scotland and were brought to the United States in 1822. Ayrshires are red and white. The red can be any shade, sometimes quite dark. The spots are usually jagged at the edges. Cows are medium size (about 1,200 pounds) and bulls weigh about 1,800 pounds. Ayrshires are noted for good udders, long life, and hardiness. They manage well in most environments without pampering and give rich, white milk.

HOLSTEIN *(photo p. 249)*

Holsteins originated in Europe. Their ancestors were black and white cattle of two tribes who settled in the Netherlands about 2,000 years ago. These high-producing milk cows were known as Holstein-Friesian and first brought to America in 1852. Today's Holsteins are black and white or red and white. The cows weigh about 1,500 pounds, and bulls, 2,000 pounds. Calves weigh about 90 pounds at birth. The cows produce large volumes of milk, low in butterfat. There are more Holsteins in America than any other type of dairy cattle.

JERSEY *(photo p. 250)*

Jerseys originated on Jersey Island in the English Channel. The first Jerseys were brought to the United States in 1850. Jerseys are fawn or cream, mouse gray, brown, or black, with or without white markings. Tail, muzzle, and tongue are usually black. Cows weigh 900 to 1,000 pounds, and bulls, about 1,500 pounds. Jerseys produce more milk per pound of body weight than any other breed, and their milk is the richest in butterfat.

BROWN SWISS *(photo p. 248)*

This breed originated in Switzerland and were first brought to America in 1869. Because of their good disposition, they are a favorite for 4-H and youth projects. Brown Swiss are light or dark brown, or gray. Cows weigh about 1,400 pounds, and bulls, about 1,900 pounds. Cows are noted for long life, sturdy ruggedness, and milk with high butterfat and protein content.

MILKING SHORTHORN *(photo p. 250)*

These cattle came from northeastern England and were first brought to the United States in 1783. Most of the early settlers had Shorthorns. They are red and white, white, or roan (a mix of red and white hairs). Cows weigh 1,400 to 1,600 pounds; bulls weigh 2,000 pounds or more. They are versatile cattle, first used for meat and milk. They are noted for long life and easy calving. Their milk is richer than that of Holsteins but not as high in butterfat as Jerseys or Guernseys.

MILKING DEVON

In 1623, two heifers and a bull were shipped to Plymouth Colony from Devonshire, England, as draft animals. The American Devon developed as a multipurpose breed. These medium-size cattle are sometimes used by small farmers for meat and milk because they are docile, easy to handle, and don't require much feed. They don't produce a lot of milk, but work well as family milk cows or on a small dairy where cattle must manage under minimal management and a high forage diet. The cattle are red, with shades varying from deep red to light red.

RED AND WHITE

This relatively new breed (the herd book was established in 1964) is a composite of several breeds utilizing red genetics. The registry began at the University of Minnesota in 1962 with an experimental herd of Milking Shorthorns crossed with red Holsteins. The breed association added other red and red-and-white dairy cattle. Other approved breeds include Ayrshire, Brown Swiss, Guernsey, Jersey, and a few dual-purpose breeds such as Normande, Gelbvieh, and Simmental.

Registering Dairy Cattle

If you buy a dairy cow or heifer that is already registered, check with the breed association about membership. Belonging to a breed association has advantages. The association can provide you with educational material to help you get started breeding dairy animals and can also help you market your heifers later if you decide to sell some.

Register or transfer the registration of your new animals to your name. Be sure the color markings or the tattoos listed on the registration certificates are correct. For Ayrshire, Guernsey, and Holstein breeds, you may use photos of both sides of the calf or heifer, or sketches of her markings. For the solid color breeds such as Jersey and Brown Swiss, you need an ear tattoo.

Buying a Cow

If you buy a mature cow, she should already have had a calf. You can judge her udder and conformation fairly easily. You need to know several other things about her, however, such as whether she's had her vaccinations, whether she has been tested for brucellosis and tuberculosis, and whether she is pregnant.

If your cow is dry and about to calve, you know she will soon be milking again. If she is milking now, make sure she is pregnant again for her next calf. Unless she calved within the last few weeks, she should already be bred again. If she has been milking several months and is not pregnant, she may have a fertility problem.

Where to Buy a Cow

You may be able to buy a cow or several cows from a neighbor or a local dairy. Check the ads in your regional farm newspapers, ask at the feed store, or see if your local veterinarian or county Extension agent knows of anyone with cows for sale.

The best place to buy cows is from a dairy rather than the auction yard. Most cows going through auctions are being culled for one reason or another. They may be old or crippled, no longer fertile (unable to become pregnant), or have some other problem you don't want. There may be a few good ones among them, but you can't always tell which ones they are. For example, a nice-looking young cow may be infertile because of a uterine infection.

If you go to a dairy, you may be able to purchase a young, healthy cow (with a long life ahead of her) that is not producing enough milk to pay her way in a large commercial dairy. She may be fine for your purposes, however. A family milk cow, or a cow in a small dairy operation, doesn't need to be a world-beater with top production. The dairyman may sell her to you instead of sending her to auction. If he is willing to do that, ask him how much extra it would cost to have her bred before you take her home.

The advantage to buying directly from the owner is that you can get the cow's history and know what vaccinations she has had. You need to know if she has been checked for Bang's disease and TB (tuberculosis). In some states, every cow that produces milk for human consumption must be certified free of these two diseases.

Another place to buy a cow is a dispersion sale or farm sale (when a dairy is selling out), a production sale where a breeder is selling springer heifers and a few cows, or a consignment sale where several registered breeders are offering cattle. The latter will be the most expensive kind to buy.

How to Choose

If you've never had a milk cow before, try to choose a gentle cow that's had a calf and is milking, but not at peak lactation. If she's nearing the end of her milk production cycle (ready to dry up in a couple months, in preparation for her next calf), she'll be easier for you to learn to milk. It won't be so hard on her if you don't get her completely milked out the first times you milk her. Your first cow might be an average producer that does not need large amounts of high-energy feed. She can do well for you and keep her body condition without becoming thin.

If you want a cow that gives rich milk — lots of cream for butter and ice cream — choose a breed that has milk with a high butterfat content. If you are more interested in high volume or low-fat milk for drinking, choose a Holstein.

CONFORMATION

The mature cow is easier to judge than a heifer or young calf. What you see is what you get. With a younger animal, you need a little imagination to determine what she will look like when fully mature.

The dairy cow's head and neck should be long and slim, with a top line that is slightly concave going to the withers. Her back should be straight from shoulders to tail. The line from her hips to her pin bones (the protruding bones on either side of the tail) should be level or only slightly slanted downward toward the rear. If it is slanted upward, she may have calving problems.

Her feet and legs should be relatively straight when viewed from front or behind. Her shoulders should not stick out from her body. From the side, her hind legs should not be sickle hocked (with too much bend at the hock, putting her feet too far forward under her body). She should have strong, upright pasterns rather than weak ones in which the dewclaws touch the ground.

UDDER

One of the most important aspects of a dairy cow's structure is her udder. It should be well formed and balanced with the four quarters at the same level. The front quarters should not be higher or lower than the rear ones. The bottom of the udder (between the teats) should be flat rather than bulging.

A large udder does not necessarily mean the cow gives a lot of milk; it may be saggy with weak attachments (muscles and ligaments that hold the udder against the abdomen). You want a cow that will hold up for a long life of milking. For that she needs a tighter, stronger udder that does not sag. Looking from behind, the udder should seem well attached beneath the thighs, rather than sagging.

A pendulous udder with teats close to the ground is easily injured. The cow may develop mastitis (inflammation of the udder), for instance, just from the bruising that occurs when she bumps it with her legs while walking.

Teats should be evenly spaced, all the same size in length and diameter, and of proper size to fit comfortably in your hand. Teats that are fat or excessively short can be very difficult to grasp and therefore very hard to milk. Make sure all the teats work properly (squeeze a little milk out of each one, using your thumb and forefinger) and the milk is normal.

DISPOSITION

A cow that already has had calves and has been handled and milked, will give an indication whether she has a good disposition or a poor one. By contrast, a young heifer may be flighty and nervous just because she has not been handled as much; she may settle down and become gentle with proper handling. If a mature cow is nervous, touchy about having her udder handled, or inclined to kick, she may have been mishandled, or she may have a poor temperament. In either case, this is good reason to not buy her.

HORNS

Some cows are genetically polled (hornless) and some have been dehorned at birth. You may see some that still have horns. If you have a choice between a cow with horns and a cow without, it's best to choose one with no horns. She will be safer to handle and won't be so mean to her herdmates. She also won't be using her horns to tear up fences or get into other kinds of trouble.

Selecting Dairy Heifers

The heifers you buy as baby calves, weanlings, or yearlings, can be the foundation of a future dairy herd. Good heifers, well raised, will become good cows. Decide whether you want to raise dairy heifers to sell or to keep as cows. The best market for dairy heifers, if you plan to sell them, is for springers, which are two-year-olds about to have their first calves. Many dairies buy their future cows as pregnant heifers and pay high prices for good ones.

Some dairies will sell heifers from some of their best cows, with the understanding they can buy the heifers back from you after you raise them. If you don't have facilities for a dairy of your own, you can make a good business raising heifers.

CONFORMATION

When selecting a dairy heifer, look closely at the way she is built (conformation) and at the functional traits that help determine whether she will be a good cow. She should have outstanding breed character, which means that she closely resembles her ideal "breed type."

Choose an alert heifer with good length of body. She should have a deep, wide ribcage, a long and graceful neck, sharp (not beefy) withers, and a straight back (not humped up nor swayed down), with strong, wide loins. Like a good cow, her rump should be level and square, not tipped up at the rear, nor slanted downward. If her rump is tipped up and her tail head too high, she will have more trouble calving; the angle of her pelvis will make it hard for the calf to come up over it and enter the birth canal.

Her hind legs should be relatively straight when viewed from behind — neither too close together at the hocks nor splay footed (toes pointing out) — and set squarely under her body. Front legs should be straight. She should walk smoothly, without throwing her feet out to the side or swinging them inward. Avoid heifers with coarse, flat-topped shoulders or low, saggy backs. A heifer should be well balanced and well proportioned in all of her body parts. She should not be short-bodied, shallow-bodied, or too short-legged. She won't "grow out of" this type of poor conformation.

UDDER

Teats should be well spaced, just as for a mature cow. Sometimes the teats of a heifer (or one that has just calved) are too short and small for comfortable milking, but they usually become a little longer after the young cow has been milking awhile.

Buying Calves

Baby calves are fun, and easier to tame and handle than an older calf or yearling. You can gentle the calf and earn its trust before it gets bigger and stronger than you are. Baby calves are cheaper to buy than older dairy heifers, but they also are more of a gamble because they may get sick. You must spend a lot of time caring for them and be the substitute "mother" for a while. They can be a delightful way to get started with dairy cattle.

Once you get calves weaned off milk, they are easy to care for by raising them on good pasture with a little hay and grain. Good dairy heifers are worth more money than beef heifers, since a good dairy cow makes more money producing milk than a beef cow can make by producing beef calves. If you sell dairy heifers as yearlings or breed them AI and sell them as springers, ready to calve, they can be a very profitable venture.

Most dairy farms have young calves to sell, especially in the spring, since few dairies raise their own replacement heifers. They are usually too busy with the milk cows to take time to raise heifers, so it's easier to just sell the calves, let someone else raise them, then buy pregnant heifers whenever they need to replace the older cows that must be sold.

Dairy cows must have a calf every year to produce their maximum amount of milk; cows produce a great deal of milk after they "freshen" (give birth and start producing milk). A milk cow's peak production occurs a month or two after calving. From then on, her volume of milk gradually declines. To have maximum production, dairy cows are bred every year to have new calves and are allowed to "dry up" briefly before the new calves arrive. This gives them a rest from milking and a chance to prepare for having another calf.

Some dairies sell all of their newborn calves. Others keep the heifers to raise, keeping some and selling springers to other dairies. Bull calves are almost always sold as soon as they are born and are cheaper than heifers. A few calves might be crossbred (half beef) if the dairyman breeds his heifers to a beef bull that sires small calves for easy birth. The crossbred calves can often be purchased quite cheaply. The steers may make good bucket calves to raise for beef or 4-H projects. A crossbred heifer might become a family milk cow or nurse cow, but she will not give enough milk to be a good dairy cow.

Some dairies take their newborn calves to a nearby auction to sell them. In areas of the country where there are lots of dairies, there are always baby calves at auction sales. This is the riskiest place to buy a young calf. Even though the calf may have been healthy when taken to the auction, it may become sick after you bring it home. This is because some calves are taken from their mothers and sold before they have a chance to nurse colostrum. They do not have the antibodies they need to protect them from diseases.

A sale yard is a likely place to pick up diseases. Cattle come and go and spend time in pens before being sold. Some of them may be sick or coming down with an illness. Even though most of the cattle that go through the pens are healthy, there's a chance for disease germs to contaminate the pens. A young calf with low immunity can pick up disease during the time it spends in a pen, waiting to be sold. Some calves will become very sick soon after you bring them home and may be difficult to save.

Don't buy a calf at an auction, if you have other options. Check the local dairies. Find out which ones would let you pick out a new calf or several calves. If you buy calves at a dairy, you can ask questions and find out more about the calves. This is especially important if you plan to raise heifers. You'll want to know something about the sire and dam, if possible. And you can make sure you only buy calves that have had colostrum.

Evaluate the Calf

If you are buying a heifer calf to keep or sell as a replacement heifer, pay close attention to her conformation. It may be harder with a baby calf than with a larger heifer to tell what she'll look like when she grows up, but there are still clues. Pay close attention to her legs and body. Her feet and legs should be relatively straight and not crooked, and her body shape should be like that of an older heifer.

Future teat size and length can be determined by the size and shape of her little udder. You don't want a heifer with long teats or fat teats. Actual shape and size of the udder itself when she grows up will be difficult to determine, however, so it helps if you can see what her mother's udder looks like (which you can do if you purchase the calf at a dairy as a newborn).

If you buy bull calves to raise as beef animals, conformation is not quite as important. You will castrate them and raise them until they are big enough to butcher or sell as good-sized yearlings to go into a feedlot. They don't have to be as structurally perfect as heifers that must have a long and productive life. But make sure they are good-looking calves because you want them to do well; a set of uniform, nice-looking dairy steers will sell better than crooked-legged, sway-backed animals.

If you plan on raising dairy steers for beef animals, choose a breed that is larger-bodied; Holsteins, Brown Swiss or Shorthorns are better than Jerseys or Guernseys — especially if you hope to sell the steers later. Holstein or Brown Swiss steers bring a better price per pound than Jersey steers, for instance.

AGE

A newborn calf is always cheaper but more risky to buy than an older calf. Newborn calves are more likely to get sick unless you buy them from a well-managed dairy and take very good care of them through their first weeks of life. The future conformation of a heifer calf is also more difficult to predict in a newborn calf than in a started or weaned calf.

A started calf — one that is several months old — is easier to judge and less apt to become ill. The easiest calf to care for is three or four months old and already weaned. It no longer needs milk replacer (bottle or bucket) and is past the critical age for scours. Started calves cost more.

CALF HEALTH

Be sure calves are healthy before you bring them home. A calf should look bright and perky, lively and energetic, with glossy-hair coat and sparkle in her eyes. Bowel movements should be firm but soft, neither hard pebbles nor runny. If the calf is very young, make sure her navel stump has dried. If the navel is still moist, the calf may be at risk for navel infection. Be cautious about buying a calf with a moist navel stump; this may mean it was not disinfected with iodine.

Check the color of the calf's feces. If the newborn calf is still passing dark-colored bowel movements, she may not have had any colostrum. Once a calf has nursed, the colostrum begins to come through and bowel movements are a bright yellow color.

If the calf is dull or slow moving, has a dull or rough hair coat, foul-smelling manure or droopy ears, she is sick. Other signs of sickness are cough, runny or snotty nose, or the calf standing with her back humped up. If you are in doubt about the health of a calf you are considering, have someone else look at her, too. An experienced person is always a help when selecting dairy calves. Don't be in too big a hurry to buy calves. Look at several and ask questions so you can make a wise choice.

CHAPTER TWELVE

Care of the Dairy Heifer

B EFORE YOU BRING A NEW DAIRY CALF HOME, make sure you have a good place for her. All calves need shelter, but the brand-new calf is the most fragile. She needs to be kept warm and dry.

Fix a Place for the Calf

One of the best ways to prevent stress and disease is to have a good place for the calf, with adequate shelter that is clean and dry. In winter she needs protection from wind and cold. In summer she needs shade and good ventilation to avoid heat stress. In a mild climate a calf may need only some clean bedding in a protected fence corner with boards or plywood on the sides for a windbreak and roof over it. But if weather is cold, she'll need a warm barn stall or even a place in the garage or back porch for a while, until she is several days old and able to live in an outdoor shed.

When transporting the calf home, make sure she does not get chilled during the trip, or she may get pneumonia. In warm weather, she can ride in a pickup with a rack. If it's cold, rainy, or windy, bring her home in a van, camper shell, or enclosed trailer.

A protected fence corner is adequate shelter in mild weather.

The calf should always have dry, clean bedding. Moist, dirty bedding contains harmful bacteria. If she was sold soon after birth, her navel may not be completely dry yet; she could be at risk for navel ill if she lies in a dirty place (see page 127). If you buy newborn bull calves to raise for beef, this can be a very important issue, since their navel stumps take longer to dry up.

Damp bedding also conducts warmth away from the calf's body and she may chill. Ammonia gases put off by bedding that is wet from urine and manure can irritate and weaken the calf's lungs and allow bacteria to become established, leading to pneumonia.

Keep each calf in a pen by herself to prevent the spread of disease and to keep calves from sucking on one another. Individual calf hutches are ideal. The hutch is a small shed with an attached outside pen; each calf has her own little barn and a yard next to it for exercise and sunshine. Be sure it is well ventilated so the hutch won't become a miniature oven on hot days.

Calf hutches can be about 4 feet by 6 feet or a 4-by-8 board pen with a roof over one end. Hang a water bucket and feed tub from the wall about 20 inches off the floor or ground to keep the calf from stepping in them or getting manure in them.

Keep hutches very clean. Thoroughly clean and disinfect them between calves. Get rid of all old bedding and scrub the walls and floor with a good disinfectant. Your veterinarian can recommend a good product to use.

Feeding the Newborn Calf

Buying a calf from a local dairy gives you the advantage of being able to take good care of her from the start; you can take her home directly after she is taken away from her mother. She does not have to spend time without food at an auction yard.

The calf should have had a nursing of colostrum as soon as she's born. Some dairymen let calves stay with their mothers until they nurse once or twice. Others prefer to take a calf away before it has nursed, put it in a clean pen, and feed with a bottle. Colostrum from the cow is milked out and saved for feeding calves.

When bringing a new calf home from a dairy, ask if you can buy a gallon or two of fresh colostrum to take home with you. Keep the extra colostrum in your refrigerator and feed it as long as it lasts. Use very clean containers for storing it. Before it's all gone, start mixing some milk replacer with each feeding to gradually accustom the calf to the new taste.

Many dairymen who raise their own heifers prefer to feed the calf by hand, teaching her to drink from a bottle or bucket from the start. Then they know how much colostrum the calf actually drinks. On the first feeding, let the calf drink as much as possible so she can take in antibodies to protect against disease. After that first feeding, though, don't let her overeat. A dairy cow gives so much milk that a greedy calf may drink too much and end up with diarrhea.

Baby Bottles

Teach the calf to nurse a bottle. If the calf was with her mother a while before she was sold, she knows how to nurse from a cow but not a bottle. Your first task is to quickly teach her to nurse a bottle. You don't want her to go hungry very long. A young calf should be fed several times a day.

It's easier to teach a calf to drink from a bottle if she has never nursed its mother. A newborn calf is hungry and will eagerly suck a bottle for her first meal. A calf that has already nursed her mother prefers the taste and feel of the udder. She wants Mama! These calves

WHAT TO DO WITH EXTRA COLOSTRUM

You can save money on milk replacer by feeding the extra colostrum and transitional milk from your cows. Colostrum and transitional milk cannot be sold as milk, but they make good food for calves. Every time a cow calves, her milk for the first five days can be used to feed all your calves, or to mix with milk replacer. The 8 to 25 gallons of colostrum/transitional milk produced by a cow should not be wasted.

All that colostrum may not fit in your refrigerator. It can be stored at room temperature as sour (fermented) milk. Don't feed sour milk to a calf until he is more than two days old. Fermentation reduces the antibody content of the milk, and the acid in the fermented milk may upset the newborn calf's digestion.

Store extra colostrum and transitional milk in a clean plastic garbage can or other plastic container. Keep it covered to prevent contamination with dirt or flies. Don't use metal containers; acid produced during fermentation will corrode metal. Store containers in a cool place (below 70°F), but they do not need to be refrigerated. Temperatures above 75°F will cause excessive souring, which isn't good for the calf. Properly stored colostrum will keep well for two or three weeks after it ferments. Stir it each time you add more or take some out for feeding.

Colostrum contains almost twice as much food solids as regular milk. When you feed fermented colostrum, dilute it with water (two parts colostrum, one part water). If it is mostly transitional milk, it won't need to be diluted.

can be stubborn, so it takes patience to start them nursing a bottle, especially if it isn't colostrum; the taste of the milk is different from what they nursed from mama.

To feed the calf a bottle, back her into a corner so she can't get away or wiggle around too much. Straddle her neck with your legs. This will enable you to have both hands free to handle her head and the bottle.

Use a nipple that flows freely when the calf is sucking so she won't have to work too hard at it and become discouraged. It should not flow *too* fast, or it may choke her. Hold the bottle so the milk will flow to the nipple and the calf won't be sucking air. Don't let her pull the nipple off the bottle.

If a calf's first feedings with a bottle are colostrum instead of milk replacer, she'll be more willing to suck, since she prefers the taste of colostrum. It is the best food for her at this time. Even though she is too old to absorb more antibodies directly into her bloodstream, the antibodies still have some protective effect in the gut for a while. Colostrum helps combat germs that cause scours and has a laxative effect to stimulate her to pass her first bowel movements.

If there is no way to obtain colostrum, use whole milk, preferably raw milk from a dairy instead of pasteurized milk from a store. Calves like the taste of whole milk better than the taste of milk replacer and will drink it better than milk replacer while teaching them to nurse a bottle. Whole milk is the best feed for a calf. The biggest benefit of milk replacer is its price (compared to whole milk) and ease of storage.

Before you run out of colostrum or whole milk, start mixing it with milk replacer if that's what you'll be using to raise the calf. The

IS THIS MILK SAFE FOR CALVES?

Milk from a cow with mastitis can usually be fed to calves, if the milk is not abnormal. Calves can also drink milk from any cow being treated with antibiotics, even though the milk can't be used by humans until the withdrawal time for that antibiotic is past. Mastitis germs usually won't make a calf sick because they are different from germs that cause gut infections. If you feed milk from a cow that can't be used by humans, feed it only to calves penned individually, so they can't suck on each other. Mastitis germs can be passed from a calf's mouth to another calf's udder if they suck on each other. To be safe, it's best to pasteurize any mastitis milk fed to calves.

calf can gradually adjust to the taste of what she will be drinking from now on. If you make the mistake of suddenly switching to milk replacer, she may hate the taste and refuse it.

Many combinations can be used for feeding calves, including mixes of whole milk and milk replacer, or sour colostrum and milk replacer. Just be sure to make any changes gradually

Bottle feeding a newborn calf.

to avoid digestive upsets. Don't suddenly switch a calf from milk replacer to sour colostrum. Mix it. It's best if you do not change the feed of a young calf during her first two weeks of life. After that, the changes won't affect her much if they are done gradually.

Keys to getting a stubborn calf to suck a bottle are patience, persistence, and real milk or colostrum that has been warmed. Young calves hate cold milk. Warm the milk enough that it feels pleasantly warm on your skin but not hot. If it's too hot, it will burn the calf's tender mouth and she won't suck. Never overheat milk or milk replacer. Overheating damages proteins and vitamins. An older calf doesn't need milk to be warm, but still prefers warm milk to cold.

Keep feeding equipment clean. Always wash the bottle or nipple bucket after each feeding, or bacteria will grow on it and may make the calf sick. Nipple buckets must be taken apart and cleaned. Use a bottlebrush to thoroughly clean a bottle.

Feeding Schedule

For the first few days it's best to split the daily feeding into three portions and feed every 8 hours. Feed the calf early in the morning when you first get up, again in the middle of the day (early afternoon), and then the last thing at night before you go to bed. Once the calf is a week old, it can be fed twice a day (every 12 hours, morning and evening).

Don't overfeed. Too much milk can upset her digestion and cause diarrhea. Feed the calf according to her size. A big calf needs more than a small one. A newborn Holstein may weigh 90 to 100 pounds, whereas a newborn Jersey may weigh only 50 pounds. Weigh or measure the milk to make sure you are not overfeeding or underfeeding her.

Mix the milk replacer with warm water according to directions. Feed 1 pound (about a pint) of milk daily for each 10 pounds of body weight. A calf that weighs 90 pounds needs a total of 9 pints daily. That would be 4½ pints (slightly more than 2 quarts) in the morning and again in the evening, or about 1 gallon per day.

A 50-pound calf needs only 5 pints per day; each feeding would be a little more than a quart. A 75-pound calf needs 7½ pints daily (a little less than a gallon); each feeding would be about 1¾ quart. If you are feeding a very young calf three times a day, split the daily amount into three equal portions.

CALF FEEDING PROGRAM

AGE	RATION
Birth to 3 days	colostrum
4 days to 3 weeks	whole milk or replacer; grain mix or starter.
3 to 8 weeks	whole milk or replacer; grain mix or starter, with access to good roughage.
8 weeks to 4 months	2 to 5 pounds of calf ration (grain mix) with access to good roughage; calves can be weaned as early as 8 weeks but do better if weaned a little later.
4 to 12 months	3 to 5 pounds of calf ration with access to good roughage.

A nipple bucket can simplify your feeding task.

Feed the same time each day, on a regular schedule, and you won't upset the calf's digestive system. If the calf starts to get diarrhea from being overfed, immediately cut the amount of milk in half for the next feeding. Then gradually increase to the recommended amount for her size. If diarrhea is due to a gut infection, immediately start the calf on electrolytes and antibiotics. As she grows, increase the amount of milk to be appropriate for her body size, but don't feed more than 12 pounds (one and a half gallons) per day.

Bucket Calf

Once your calf learns how to suck a bottle, you can teach her how to use a nipple bucket. Then you can hang the bucket from her fence or stall wall. The advantage of a nipple bucket is you don't have to hold it while the calf nurses. When feeding a group of calves, this makes your job a lot easier and saves time. Hang buckets a little higher than the calves' heads, so they can reach them easily.

If you use a nipple bucket, don't enlarge the nipple hole. Some people widen it so milk flows faster, to decrease the time it takes the calf to drink the milk. When milk runs too fast, the calf may breathe some of it "down the wrong pipe" because she can't swallow fast enough. This will make her cough, and could lead to aspiration pneumonia, caused by milk getting into the lungs. The irritation and infection that results may kill the calf.

Teach a calf to drink from a pail by having her suck your fingers and then immersing them in the milk while she's sucking on them.

You can teach your calf to drink from an open pail if you wish. Put fresh warm milk in a clean pail and back the calf into a corner. Straddle her neck and put two fingers into her mouth. While she is sucking your fingers, gently push her head down so her mouth goes into the milk. Spread your fingers so milk goes into the calf's mouth as she sucks. After several swallows, remove your fingers. Repeat this as often as necessary until she figures out she can suck up the milk. A pail is easier to wash than a nipple bucket or bottle. It's easier to teach a calf to drink from a bucket if she has never sucked a nipple (calves prefer the nipple and may refuse to drink from an open pail).

Milk Replacer

You can buy milk replacer at feed stores. There are many different kinds and brands, and some are better than others. Read the label on the bag to know what a certain product contains. If it's confusing, ask a dairyman or your county Extension agent for advice on recommending a good brand for a young calf.

Avoid scours. Diarrhea in baby calves is not always due to infection. It can also be caused by nutritional problems such as overfeeding or use of poor-quality milk replacers.

Protein and fat content: The National Research Council (NRC) recommends using milk replacer with a minimum of 22 percent protein and 10 percent fat. Young calves do better if it contains more fat than that. Milk replacers with 15 to 20 percent fat are better; the calf will grow faster and be less apt to get scours from inadequate nutrition.

Fiber content: Check the fiber level in the milk replacer. Low fiber (0.5 percent or less) means it has more high-quality milk products and not so much filler.

Protein sources: See whether the protein sources are milk based or vegetable protein. Milk protein is the highest quality and best for the calf. This is because the newborn calf has a simple stomach. Her rumen is not yet able to digest fiber and roughages. She can digest and use protein from milk or milk by-products more easily and efficiently than protein from plants.

To mix milk replacer, follow directions on the bag. The powder is mixed with warm water and fed like milk. The recommended amount varies with different breeds. The powder will mix better if you put warm water in your container first, then add the powder and stir well, until it is all dissolved. It won't mix as quickly if water is cool; have the water a little hotter than for feeding the calf, just enough so it will be the right temperature by the time you mix in the powder and take it out to feed the calf.

Store milk replacer properly. Keep powdered milk replacers dry and clean; the powder spoils if it gets damp. Close the bag again after each use. Keep it in a container with a tight cover. Quality may be reduced and it can become contaminated with germs if the bag is left open and exposed to light, moisture, flies, and mice.

Combating Diarrhea

If the calf gets sick, she may be reluctant to nurse the bottle and may have runny manure. If too much milk causes the diarrhea, drastically reduce the amount of the next feeding. If the diarrhea is caused by infection, the calf will need immediate treatment.

Medications for scours are covered in chapter 10. Make sure the calf gets extra fluids. You may be able to give warm water and electrolytes with a nursing bottle if she will drink it, but most sick calves will refuse. You'll have to give fluids with an esophageal feeder.

The most damaging effect of diarrhea is loss of water and important body salts. When pathogens irritate or destroy the intestinal lining, it can no longer absorb fluids properly. The bowel movements of a calf with diarrhea may contain 5 to 10 times as much water as normal and the calf becomes dehydrated. A calf may die because of fluid loss if you don't replenish it.

Encourage the calf to nurse regular feedings of milk replacer in between fluid treatments. She needs the energy provided by milk to keep up her strength. If she is too sick to nurse, or won't nurse enough, give her the regular feedings with an esophageal feeder. Give medicated fluids and electrolytes between feedings.

Don't mix medications and milk. Do not mix electrolytes containing sodium bicarbonate with milk or milk replacer. This prevents curd formation in the stomach and can make diarrhea worse. Wait two or three hours after the milk feeding before giving fluid containing electrolytes. For a dehydrated calf the fluid/electrolyte mix should be given every six to eight hours or until she feels well enough to nurse again. The calf needs to be nursing or getting fluids at least four times a day to ward off dehydration.

Starting a Calf on Solid Food

A growing calf needs concentrates and roughages. Concentrates are feeds that are low in fiber and high in energy, such as grain. Roughage (forage) is feed that is high in fiber (bulkiness) and low in energy. Hay, grass, corn silage, straw, and cornstalks are roughages. A calf needs roughage to help develop her digestive system so her rumen can begin to function properly.

Always make sure calves have fresh, clean water every day, and access to trace mineral salt. Calves still need water, even though they are getting fluid with their milk replacer. Water is especially important in hot weather.

Grain

Teach the calf to eat grain or calf starter pellets as soon as possible. Put some in her mouth after each milk feeding until she learns to like it. Then you can feed it in a tub or feed box. At first, about one cup of grain (one-fourth pound) is all a young calf can eat each day. Increase it gradually until she is eating about two pounds (eight cups, or two quarts) of grain daily, split into two or three feedings, by the time she is three months old.

Hay and Pasture

A calf should have hay as soon as she'll start to nibble on it. Calves have small mouths and can't handle coarse hay, but will nibble on tender, leafy hay. Fine alfalfa, clover, or grass hays — or a mix of these — are very nutritious. Don't give her much hay at a time or she'll waste it; baby calves won't eat hay that has been tromped on or laid on. Give a little bit of fresh hay twice a day. Young calves generally don't overeat on alfalfa hay and there is less risk of bloat than with an older calf because the young calf has a much smaller and less functional rumen.

You may want to turn out the calf into a pasture at this point instead of keeping her in a hutch. Good green pasture is excellent feed to supplement a calf's diet, as long as she is getting milk or milk substitute, grain, and some hay. Calves that have been on hay may experience looser feces when put on green pasture, just because the grass has a higher water content.

If a calf is on pasture, she may be at risk for picking up internal parasites. In a clean hutch by herself, worms will not be a problem, but after she goes out on pasture with other cattle or where cattle have been, she may need to be dewormed. Ask your veterinarian about a deworming schedule and which products to use.

Management Procedures

Vaccinations can wait, if the calf has had adequate colostrum, for four to six weeks, when the temporary immunity from antibodies in the colostrum wears off. Ask your veterinarian about a vaccination program for your calves.

Other management procedures for very young calves include castrating a bull calf (see page 152), dehorning (see page 153), removing extra teats on a heifer, and halter training. For some of these

To flank a calf and lift him off his feet (to lay him down on his side), grasp a front leg just above the knee and the loose skin of the flank.

procedures, you want the calf restrained and still; sometimes this is easiest when the calf is lying on the ground.

To put a calf on the ground easily and gently, without a struggle, flank the calf. Stand close to the calf, then reach over its back and grab hold of the flank skin with one hand and front leg (at the knee) with your other hand. Lift the calf off its feet, gently lowering its body to the ground. To hold the calf still while it is lying on its side to be castrated or have extra teats removed, kneel down and hold the calf's front leg (folded at the knee), putting gentle pressure on its neck with your knee, so the calf cannot rise.

Removing Extra Teats

Most heifers have just four teats but some are born with an extra one or two. Even though these are usually small and do not develop, extra teats are of no use to a cow and may cause problems when she is milked. They also may leave her vulnerable to mastitis. If a calf has extra teats, they should be removed.

As soon as a heifer is big enough to tell which teats are the extra ones, they can be removed, generally when she is one to three weeks old. If you're not sure which one(s) should be removed, wait until she is older and teats are more developed and have your veterinarian do it. Later teat removal induces more bleeding, and the wound may need stitches.

Extra teats on a baby heifer are easily removed. Flank the heifer and put her on the ground. Extra teats are easier to locate when she's lying down; you can examine her udder closely. The four regular teats are arranged symmetrically with the two rear teats slightly closer together than the two fronts. An extra teat is usually smaller and located close to the main teats.

After determining which one to remove, disinfect the teat and snip it off with sharp, disinfected scissors. Hold the scissors with the handle toward the front of the calf and blades pointing toward her rear. Make the cut lengthwise with the body. Then dab the severed area with iodine.

Remove the extra teat with clean scissors, cutting lengthwise along the body.

Another method is to tie the heifer and remove the extra teat while she is standing. It helps if someone holds her back end so she can't move around. Pull the extra teat down and snip it cleanly off where it meets the udder, making the cut lengthwise with the body. Swab with disinfectant. In a young heifer there is rarely any bleeding. Use fly repellent if it's summertime.

Halter Training

Put a small halter on your heifer calf and teach her to lead and tie up. This is the easiest age to halter break, since she is not bigger and stronger than you are. A halter-trained heifer will be easier to handle and work with the rest of her life.

Don't spoil a heifer. It's all right to make a pet of her, but don't let her get away with unacceptable actions like butting, kicking, or dragging you around when you try to lead her. She must respect you and learn to behave around people. If she is handled in a gentle but firm manner and not allowed to become pushy, she will grow up with a good attitude and be a pleasure to own.

When halter training a young calf, make sure the halter fits (so it won't pull off), and keep the lessons short at first. When leading, use a pull-and-release technique rather than a steady pull. If she comes a few steps as you pull, reward her by slacking off the pressure. When she learns that the pressure eases up when she walks forward, she will lead. Walk on her left side, holding the rope in your right hand, about a foot from the halter.

When tying, tie to something sturdy, at calf's eye level or higher, with about 12 inches of rope. You don't want so much slack in the rope that she'll get her foot over it. As soon as she quits fighting the rope and stands patiently, turn her loose. As she gets used to being tied, leave her tied a little longer. Once she accepts the restraint of a halter, you merely need a few refresher lessons now and then, and she will stay halter trained the rest of her life. When she grows up you can tie her whenever you need to restrain her for something, or to teach her how to be milked.

Weaning

A dairy calf can be weaned off milk or milk replacer as young as eight weeks of age. Dairymen often wean calves this early because milk or milk replacer is more expensive than grain and hay. A young calf weaned too early, before she's eating enough grain and hay, will not grow as well; it's better if she nurses longer. Weaning age depends on available feed sources, heifer health, and how long she has been eating solid food. Most dairy calves do better if they are not weaned until they are about three months old.

Calves can be weaned when they are young if they are eating a special high-quality dry starter ration containing milk products. If the calf eats enough of this starter, she won't need milk and early weaning can reduce your costs. The starter ration is less expensive than milk replacer. Do not wean her until she is eating about two pounds per day of a grain concentrate.

Start feeding a dry feed ration as soon as possible. The grain starter containing milk products can be offered as early as the first week of life, and grain should be offered by three weeks of age. She won't consume very much at first, but she needs to learn how to eat it. Give her all the grain starter she'll clean up at each feeding; cut back the amount if she wastes some. By the time she is three to four months old, she may be eating as much as four to five pounds of grain each day. She should also be getting some good hay to help develop the rumen. Very young calves don't have a working rumen; their digestion, like a human's, is done with a simple stomach. They can digest milk, but not roughages. They must learn to eat roughages and start "gut bugs" breaking down and digesting roughage in the rumen.

Calves on a complete starter ration containing grain and roughages can be fed as much as they want; leave feed available at all times. Start feeding hay at about three months of age. Grain starter or a complete starter ration is useful calf feed until four months of age; it can help her through the weaning process. Calves should be eating at least one pound of starter daily for every 100 pounds of body weight before they are weaned. Use a weight tape to estimate the calf's weight.

Overlap hay and starter ration. Before you discontinue the starter, give hay for at least two or three weeks. Make sure she's eating hay before you switch from starter to grain. Calves on grain starter should always have hay in addition and should be eating hay for at least a week before weaning.

The Weaning Process

When you wean a calf, it's easier on her if it's gradual. Start by decreasing the amount of the twice daily feeding of milk replacer. Cut back to about three-fourths of what you've been giving her. Do that for a few days and encourage her to eat more grain, feeding it after she finishes her bottle or bucket. She'll then be interested in the grain and not so mad at you for shortchanging her on milk. Decrease to one feeding of milk per day, giving grain at her other feeding time. When you stop the milk feedings entirely, give grain at the times you used to feed milk.

If you started her on hay at an early age, the rumen has started to function and enlarge. Baby beef calves start eating hay or grass at a few days of age, following their mother's example. But the dairy calf does not have mama to copy — no one to show her how to eat. You must be her substitute mother, encouraging her to eat hay and grain by putting some in her mouth.

It takes time for the rumen to enlarge to accommodate solid food that will give a calf the nutrition she needs. The sooner she starts eating solid food, the more chance the rumen has to enlarge and develop. Young calves still don't have much rumen capacity right after weaning. They will eat just a small amount of hay compared to the amount of grain they can handle. This will change as soon as the rumen develops.

Keep some hay in front of her all the time, in a place she can easily reach. Hay should be fine-stemmed and leafy, without mold or dust. Give fresh hay often; she will be tempted to eat some each time. Feeding a large amount all at once will waste it.

If silage (good quality corn silage, or haylage, made from alfalfa/grass hay) is available, it can be fed to calves in place of some of the hay. Silage, especially corn silage, is usually not fed to young calves since it's hard to keep silage fresh and palatable. On most farms, dry hay is a more covenient source of forage for young calves, and safer because it is less apt to mold. If you do feed silage, wait until the calf is several weeks old and use only high-quality silage, daily removing any that is leftover and uneaten. Definitely remove any that begins to heat or mold. After weaning, silage is more commonly used than hay.

Keep calves separate until after weaning. If you have more than one calf, they can live together after weaning. *Don't* mingle dairy heifers before weaning. Calves like to suck on each other after their milk feedings. Even though they are getting enough nutrition from a bottle or bucket feeding, they drink fast and want more. Right after you take the bottle away or the bucket is empty, calves want to keep sucking. They turn to each other and suck on each other's ears, or even on another calf's udder.

There are problems with calves sucking each other. In cold weather their wet ears (from being sucked by a companion) may freeze. Then the ends of the ears fall off. Even if weather is not cold, you don't want calves sucking each other's udders. This can damage the tiny teats and introduce bacteria (from the calf's mouth) into the teat, causing infection that could ruin that quarter of the udder. When the heifer grows up and starts to produce milk, that quarter may be non-productive.

You can put heifers together (or heifers and steers) about two weeks after weaning, when they are no longer interested in nursing one another. If one of them starts sucking the other heifers, that calf should be put by itself awhile longer.

Weaning to Six Months

After weaning, young heifers can live in small groups, no more than five to a pen, or out in a good pasture with shelter and shade.

A clean pasture is the best place for heifers, giving them a little grain and hay as supplement; their rumens are still not large enough to accommodate enough forage for the total nutrition they need. If weaned calves are in pens, keep the pens clean and well bedded. It is very important that heifers do not lie on dirty bedding or in mud and manure.

Heifer mastitis and "blind quarters" can result from bacteria entering a teat when heifers lie in dirty places. A blind quarter is one that does not produce milk because it has been damaged by infection. Don't let this type of infection go untreated, or it may ruin the heifer for milking.

After weaning, gradually change heifers from a starter ration to a growing ration, which should contain at least 15 to 18 percent protein. Often the needed protein can be supplied by good pasture or alfalfa hay, with a little grain to supplement it — no more than four to five pounds of grain per day.

This is the age to make sure heifers have their important vaccinations, which should include Bang's. By now they should be dehorned and extra teats removed, if this was not done earlier. It's best to accomplish these tasks before heifers get any bigger.

You can use a weight tape to estimate a heifer's weight.

Six Months to Yearling

This is a relatively easy age, especially if heifers can be on pasture, eating just a little grain and hay as supplement. A yearling heifer won't need much pampering; a three-sided shed with dry bedding is usually adequate for shelter. She may grow a long coat of hair and look rough during winter, but if she is well fed she will stay healthy and keep growing.

Good pasture is excellent feed for heifers larger than 500 pounds. Once they are this size, you can cut back on the hay or grain feeding. Until they reach that size, younger heifers may not have the rumen size yet to eat enough green grass; they still get more nutrients per pound from good hay and some grain. Give a small heifer two to four pounds of grain daily, and a mineral supplement while she is on pasture. Pasture can supply most of the feed she needs, but be sure it is good pasture. If the grass gets short or dry, feed alfalfa hay.

Don't overfeed. Although you want your heifer to grow well, don't overfeed her on grain. It may make her too fat before she is fully grown. Her actual skeletal growth may be inadequate if she is fed a diet with too much grain and not enough good-quality roughage. If you overfeed her on grain, or shortchange her on protein (not giving her access to good pasture or enough good alfalfa hay), she may have adequate weight but not adequate frame size. In other words, her weight is due to too much fat rather than body growth.

If a large-breed heifer, such as a Holstein or Brown Swiss, weighs 800 to 850 pounds as early as 11 to 12 months of age, she is too fat. She has reached breeding weight but not breeding size. She doesn't have enough height and bone growth, especially in the pelvic area, and will have trouble calving if bred at this time.

Extremely fat heifers may not become pregnant when bred. Also, their milk-producing tissue is reduced when fat is deposited in the developing udder. The fat takes up space that would have otherwise accommodated milk glands.

Don't underfeed. Lack of feed, or poor-quality feed, may stunt a heifer. An undersized or thin heifer may not breed, and may have difficulty calving if she does become pregnant. She produces less milk, and requires more feed than a normal heifer during her first milking period; she is trying to catch up to her proper size.

Use a weight tape to measure a heifer's height and weight and to monitor growth. To estimate a heifer's weight, put the tape around her body at the girth, directly behind her front legs. Make sure she is standing squarely on all four legs. Fit the tape snugly, but not tightly, without any slack . Then read the measurement that tells her weight.

If you can't find a weight tape, use a flexible cloth tape measure or a string to determine the distance around her girth. With a string, figure out the distance by marking the string, and then measure it with a yardstick. Compare this measurement with the accompanying growth chart to see if your heifer is close to what her weight should be for her age.

Dairy Heifer Growth Chart

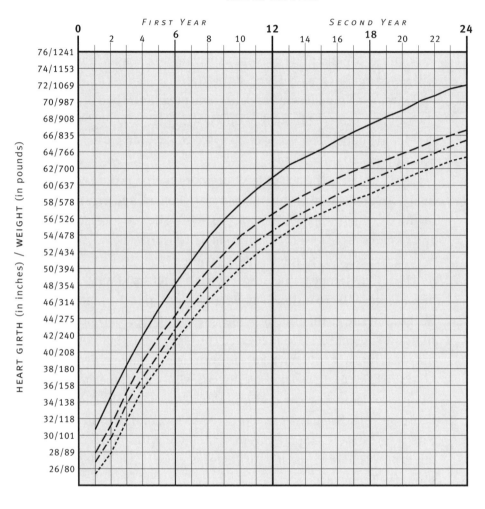

AGE IN MONTHS

FIRST YEAR

SECOND YEAR

HEART GIRTH (in inches) / WEIGHT (in pounds)

——————— Holstein and Brown Swiss
– – – – – Ayrshire
–·–·–·– Guernsey
·········· Jersey

CHAPTER THIRTEEN

Breeding and Calving the Dairy Heifer

A WELL-GROWN, HEALTHY DAIRY HEIFER can be bred at 14 to 19 months of age. The small breeds, especially Jersey heifers, reach maturity faster than larger breeds; they can be bred younger and at lighter weights than Holsteins, for instance.

Breeding Weight and Size

On average, dairy heifers are bred at 15 months of age. This is just a target goal, because skeletal size and weight are more important than age. If a heifer's frame is not large enough and her weight is too low, it is better to wait until she is of proper size for breeding. Dairy heifers should weigh 60 percent of their desired mature weights by the time they are bred. Large-breed heifers (Holsteins, Milking Shorthorns, or Brown Swiss) should weigh 800 to 875 pounds at breeding. For smaller breeds, this weight would be less. Check the chart on page 194 for the target weights and heights of various breeds. You can estimate weight with a weight tape. You can determine height by putting a flat stick across the withers and dropping a string or measuring tape to the ground from that point.

Breeding the Heifer

To freshen at 24 months, a heifer must become pregnant at 15 months. Most heifers calve about 9 months and 7 days after breeding.

Heifers give obvious signs of heat by trying to ride other cattle, standing for other cows to mount, bawling, and pacing the fence. Signs are harder to detect if there are no other cattle around. Watch for agitation, tail swishing, clear mucus discharge from the vulva that is slippery to the touch, and a red or swollen vulva. If there's a bloody tinge to the discharge, or blood on her tail or buttocks, this is usually a clue that she was very recently in heat and you missed it.

Keep records. When a heifer starts cycling, keep a record of her heat periods, which are usually three weeks apart. The pattern will help predict when she should be bred. If you know when her last heat period was, you can watch closely for the next one, which will come about 17 to 25 days later. If you keep good records, you won't be as likely to miss the time to breed her.

Breed a dairy heifer to the best bull available. If she has a heifer calf, it will be worth more if it was sired by a good bull of her breed. If you don't plan to sell the calf as a dairy cow, you might breed her to a beef bull that sires small calves. This is a way to make sure she does not have a difficult time with her first calving. A crossbred dairy/beef calf makes good beef if you want to butcher it or sell it for beef. Choose the sire ahead of time and make arrangements with the AI technician so you can purchase semen and have the heifer inseminated at the proper time.

Artificial insemination makes it possible to select an outstanding bull of any breed. These bulls are kept in "bull studs" where semen is collected and frozen. Request a catalog from your AI technician.

Once a heifer is bred, watch her closely for the next few weeks, especially during the time she would have her next heat period. If she does not come into heat at that time, she's probably pregnant. To make sure, have your veterinarian check her for pregnancy about two or three months after breeding.

GUIDELINE FOR WEIGHTS AND HEIGHTS OF DAIRY HEIFERS

(height is measured from ground to top of withers)

Age in Months	Large Breeds (Holstein, etc.)		Small Breeds (Jersey)	
	Weight (pounds)	Height (inches)	Weight (pounds)	Height (inches)
0	96	29	55	26
2	170	34	115	30
4	270	39	195	34
6	370	44	275	39
8	500	46	385	41
10	600	48	460	43
12	700	50	520	44
14	800	51	575	45
16	900	52	650	46
18	990	53	730	47
20	1,050	54	800	48
22	1,175	55	875	50
24	1,300	56	960	51

Prepare for Calving

Once your heifer is bred, you have nearly nine months to prepare for calving. A pregnant heifer needs little care except for good feed.

During pregnancy, feed her for good growth so she'll reach ideal two-year-old weight and size by the time she calves. She needs

AGE AND WEIGHT TO BREED A DAIRY HEIFER

Breed	Age in Months	Approximate Weight
Jersey	13 to 17	550
Guernsey	14 to 18	650
Ayrshire	15 to 19	700
Brown Swiss	15 to 19	800
Holstein	15 to 19	800
Milking Shorthorn	15 to 19	750

enough feed to accommodate her own growth as well as the growth of the calf inside her. If she's well grown and in good condition at calving time, she will hold up better in milk production. A thin heifer does not milk well and may have trouble rebreeding on schedule.

A Place to Calve

Make sure your heifer will calve in a clean place. If she calves when weather is nice, she can calve in a pasture, if you check on her frequently in case she needs help. The pasture should be grass — not dirt or mud — with shade if weather is hot. Don't put her in a pasture with gullies or ditches where she might get stuck. Make sure fences are strong so she can't go through them if she is restless during labor.

If the heifer will calve in a shed or barn, make sure her stall is clean. The stall should be large, with plenty of room for her to move around. If the heifer lies too close to the wall in a small stall, the calf may be jammed into the wall as he emerges. Put lime on the floor before covering it with new bedding. Lime not only helps disinfect the floor but also makes a non-slippery base. You need good footing so the calf can get up, and the heifer won't slip and injure herself. She'll be getting up and down a lot during labor, and you don't want her to injure her legs or damage her udder.

Make sure bedding is clean. Don't use wet sawdust, moldy straw, or any damp, dusty material. Wet or dirty bedding (with mold or manure) has germs that can invade the uterus or udder of the calving cow, or the newborn calf's navel.

The Milk Barn Habit

Prepare the heifer for the milking barn or shed several weeks before she is due to calve. Bring her into the barn and tie her in there to eat, or put her head into a stanchion to eat her grain. If she needs supplemental feed, this is a good place to feed it to her. She will become familiar with the place where she will later be milked.

Always be gentle, quiet, and calm when working with a heifer, so she trusts you and feels secure and at ease. Talk to her and familiarize her with the sound of your voice. Don't walk up behind her and startle her, or she may kick. Handle her a lot in the weeks before calving. It will make it easier for both of you when you are training her to be milked.

As you handle her before calving, brush her and groom her, touch her udder and teats, and wash them gently a few times. You want her used to being touched, not nervous or ticklish. Her udder may be sore when she calves, due to pressure and swelling. If you've handled her udder a lot, she might not try to kick you.

Signs

Start checking the heifer often as she gets near her due date or shows readiness to calve. She will develop a full udder a few days or weeks before calving. Other signs include relaxed muscles around her tail head and vulva, and teats filling with milk or leaking milk. Every heifer is different, so be alert, watching, and ready. If she's living with other cattle, put her in a separate pen so they can't bother her.

About five percent of calves die at birth. The calf may be in the wrong position to enter the birth canal properly, or the amnion sac may be positioned over the calf's head. The percent of loss at birth is even higher in first-calf heifers. These problems can be corrected if someone is there to help.

When the heifer goes into labor, put her in the barn stall or pen if that's where she'll calve. Monitor her to make sure the birth is progressing normally. All too often, assistance is given only after a cow or calf is in critical condition. Be there so you can give help or call for help quickly. If she's calving in a pasture during hot weather, make sure she and the newborn are in the shade. If she's too hot, she will tire more easily during labor. Heat stress can make the newborn calf more susceptible to disease. If she seems to be having problems calving in the pasture, bring her into the barn or pen where she can be tied up and checked. Help is most often needed for bull calves, because they are bigger, and for twins. For details on calving, refer to chapter 9.

Care of the Newborn Dairy Calf

Once the calf is born, the cow should get up and start licking him. Licking stimulates his circulation and encourages him to try to get up and nurse. Let her lick him dry. You can rub him with towels if the weather is cold.

Make sure the calf nurses and gets a good nursing of colostrum as soon as possible. Wash the cow's udder first, so the teats are clean. Help the calf find the teat, if necessary. If he doesn't nurse within an hour, milk out some colostrum and feed it to him. This is where handling and halter breaking pay off; the cows are gentle and easier to handle if you have to help them or the new calf.

If a calf is born when you are not there to observe, don't assume he nursed all right just because he's with the cow and up and around. It's important to make sure he gets an adequate amount of colostrum. Examine her udder; if one quarter is smaller than the others (and they were equally full before), you can assume he nursed, but sometimes you can't tell.

Without help, most calves eventually manage to nurse, but some won't if it is hard for them to latch onto the teats. The cow's udder may hang too low and he can't figure out how to bend his head down that far, or teats are too full and big. Even if a calf finally gets on a teat, it may take him too long. His gut lining is already starting to thicken after the first couple hours of life, and he can't absorb as many antibodies.

Left on their own, about 25 percent of dairy calves can't take hold of a teat by the time they are eight hours old. They need colostrum much sooner than that! Another 25 percent do not get enough colostrum. This means that only half of newborn dairy calves manage to nurse adequately without someone there to help them. A dairy cow's udder is so full and big that it's much harder for her calf to suckle a teat than it is for a beef cow's calf to nurse his mother.

If you are there when the calf is born you can help him nurse immediately, or feed him colostrum after milking some out of the cow. A 90-pound calf needs about three quarts for his first feeding. A 50-pound calf should have one and a half quarts.

Good colostrum is thick and creamy. Thin, watery colostrum is low in antibodies. If a cow doesn't have good colostrum, or it's bloody or she has mastitis, her calf should be fed colostrum from another cow or from your frozen supply.

After-Calving Care for the Cow

Offer the cow some feed and clean, lukewarm water to drink. She should eat and drink soon after calving. Most cows do fine after calving. Just keep a close watch to make sure all is well.

Give her as much good hay as she wants, but only a small amount of grain at first, until all the swelling is out of her udder. If a high-producing cow is fed a lot of grain just before or just after calving, she may have problems and more swelling in the udder.

If she won't eat or drink or seems dull, consult your veterinarian and take her temperature. She may have developed an infection in the uterus from calving, or she may be in the early stages of milk fever, a problem caused by a severe calcium imbalance in her body. In the later stages of this metabolic problem she will lie down and cannot get up. These are both very serious conditions but can be treated by your veterinarian if you call him in time.

Heifers often have a lot of hard swelling, called "cake," in the udder after calving. It takes several milkings to reduce the swelling; the udder may be sore until swelling is gone. Milk her at least twice a day, even if the calf is with her. This will help reduce pressure and swelling, or relieve a painful quarter the calf may not have nursed. He can't drink all of her milk.

Depending on how gentle the heifer is, you may or may not need to tie her or have her in a stanchion. If she is very motherly, she may stand quietly for milking if you do it as her calf is nursing. She may be more at ease if she is able to turn around and lick the calf, than if she's restrained.

A cow can be milked from either side, but it is traditional to milk on her right side. Train her to be milked on the right. If you ever sell her to someone else, they will most likely milk her on the right.

Sometimes a heifer's udder is so sore when she calves that she doesn't want you touching it. She may kick at her calf as you try to help it nurse, or kick at you. If she's kicking because her udder hurts, prevent these nuisance kicks by leaning your head into her flank as you milk. Press your head quite firmly against her to keep her from swinging that leg forward. Your head should be pressed into the area in front of her stifle joint — the big joint at the top of her leg. If you press hard with your head every time you feel her tensing up to kick, she usually cannot kick very well.

Cows can let down their milk or hold it up. There is no milk in the teat until the cow lets it down by relaxing certain muscles that keep the teat canal constricted. When a cow wants her calf to nurse, or thinks it's milking time, she lets the milk flow down from the udder into the teat. If a cow doesn't want to let down her milk, you can't collect much milk, so it's very important that your heifer be relaxed. If you are nervous, she will be nervous. Talk softly or hum a little tune to reassure a nervous heifer and keep her relaxed.

The udder will be tight and swollen the first few times you milk her after she calves. She may not give a lot of milk; swelling enlarges the udder more than milk does. With each milking, the swelling will reduce and she will have more milk. There may be a little blood in the first milking or two but this is usually nothing to worry about, and it won't hurt the calf if you feed it to him. The tissues of the udder have been under a lot of stress from the swelling, and some of the tiny blood vessels may have ruptured.

With patience, you can teach a heifer to stand quietly for milking wherever she may be — in a stanchion eating grain or out on pasture and not restrained. Feeding grain at milking time is a good way to encourage her to come for milking and to stand still while you milk her.

Don't leave the calf with the cow. Make sure he gets his first nursing, then put them in separate pens or stalls and start feeding him with a bottle or nipple bucket. If the calf is left with the cow, he may pick up disease germs if the udder is not clean. If the cow lies in manure out in the pasture or in her pen or stall, the calf may ingest a lot of bacteria when nursing.

A calf left with the cow for more than 12 hours may damage her udder by butting as he nurses. This can bruise the mammary tissue, especially if she has a large, full udder or a lot of swelling. The bruising might cause mastitis. If the heifer is worried about her calf, leave him in a pen or stall next to her so she can be near him. The first times you milk her she'll be more cooperative if you let him be with her to nurse while you are milking.

A cow can learn to stand still for milking, even if she's in a pen or pasture, if you feed her grain while you milk.

The calf can be fed for several days with the colostrum you milk out of the cow. The milk from your heifer cannot be sold until it no longer contains colostrum. You must milk her at least twice a day to collect what's left of the colostrum out of her udder and hasten production of regular milk. It takes four to seven days before the milk is ready to be used by people. You can keep that milk in the refrigerator to feed your calf, or freeze some. You might have many days' worth of milk for the calf.

True colostrum, the undiluted "first milk" from a cow, is only obtained from the very first milking or nursing. This is what the calf needs immediately after he is born. You should freeze the extra first milk that the calf cannot hold for later emergencies. The milkings after that are transitional milk, a mix of colostrum and regular milk. It is mostly colostrum at first, becoming more and more diluted by regular milk, until there is no more colostrum.

The colostrum is thicker and richer than regular milk, and usually quite yellow and sticky. It is waxy when cold because of the high concentration of fat. It won't go through a milk strainer. An easy way to tell if milk still contains colostrum is to watch how it goes through a strainer. Once the milk strains quickly, leaving no residue on the strainer, it is ready for human use.

CHAPTER FOURTEEN

Care of the Dairy Cow

AFTER YOUR HEIFER CALVES, you'll need to milk her. It helps if you know how to milk! Practice milking before she calves on a cow that's already giving milk.

Milking a Cow

Milking is easy, once you know how. Your arms may tire the first few times. If you've never milked before, visit a friend who has a milk cow and give it a try.

When milking a cow, make sure your fingernails are not too long. If you poke the cow's teat with a sharp fingernail, she won't like it! Before starting to milk, wash her udder and your hands. Brush off her udder and flanks with a soft brush if they are dirty. Clean hands and a clean udder will prevent germ contamination of the teats and the milk.

Milk from the right side of the cow. Make yourself comfortable on a stool beside the cow's udder. Hold a clean empty bucket between your legs. Start with front teats or back ones, or the two on the same side if you wish. Hold a teat in each hand and squeeze one at a time, squeezing the milk down and out through the teat opening. Begin the squeeze at the top of the teat with your uppermost finger and thumb grip. Finish the squeeze with the lower fingers. Applying pressure with your thumb and index finger keeps the milk from going back up the teat, so when you squeeze with the rest of your fingers, the milk comes down and out through the hole.

Aim the stream of milk into the bucket or it might squirt off to one side. When you become good at milking, you can direct the stream easily and can aim a squirt toward a waiting barn cat. Some cats love to catch a squirt of milk in their mouths.

After each squirt, release your grip. More milk will flow down into the teat. Keep up a nice rhythm by alternating squirts. When those quarters are empty, the teats will become soft and flat. You won't be able to squirt out any more milk. Then milk out the other two quarters. If one quarter seems to have more milk than another, it probably means that hand is not yet as strong as your other hand; you didn't get quite as much milk out with each squeeze and that quarter will take longer to empty.

Once you learn the proper squeezing motion, milking is easy. Your arms will tire and may ache if you milk very long. The more often you milk, the stronger they'll become, building up endurance and grip strength.

Some cows are easy to milk. Their teats are easy to grip and they let down their milk freely. The milk almost flows from the udder into your bucket. Other cows are hard milkers. Even if their teats are a nice size, it takes more effort to squeeze out the milk and longer to milk them. A cow with short teats may be difficult to milk because it's

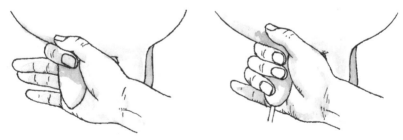

Milk by pressing with your thumb and index finger.

Squeeze with your lower fingers.

harder to get a good hold with your hand. It takes more time to milk if you can only squeeze with two fingers instead of your whole hand.

When milking, try to keep things calm and quiet so the cow can relax. Feed her grain to eat while you milk. If she is nervous or upset she will be tense and won't let her milk down. In most situations, a cow should be milked twice a day on a regular schedule, 12 hours apart. Stick to the schedule; irregular milking can be harmful to the cow.

Avoid Mastitis

The cow's udder is a complex structure that requires good care. How you handle it and the cleanliness you practice at milking time can help determine whether or not mastitis (inflammation of the udder) is a threat.

Before you start milking, check the udder to make sure there are no problems or injuries. Then check for abnormal milk. Squirt a little milk from each quarter into a small container so you can see the milk easily. If it shows any signs of mastitis (if it is lumpy, watery, bloody, or has any other abnormality), the cow must be treated.

If udder and milk are fine, wash the udder with a mix of warm water and a sanitary solution recommended by your veterinarian or a dairyman. Washing with warm water also stimulates the cow to relax the teat muscles and let down her milk. Use a clean paper towel to dry the teats before you start milking. After you are done milking, you can use a teat dip solution to help close up the teat canal and prevent bacteria from entering. To prevent sore and cracked teats in cold weather, put a little ointment like Bag Balm on the teats after each milking. This will keep the teats soft and prevent the dryness and cracking that would make them sore.

Monitor the cow's udder for mastitis. Bacteria sometimes enter the udder through the teat canal. Inside the quarter, they multiply rapidly and cause infection. The cow's body sends white blood cells to fight them, and there is often heat and swelling in that quarter. Sometimes your first clue is a stringy clot of milk when you start milking the cow, a bit of blood in the milk, or flecks or clumps that catch on the strainer. Always check the strainer carefully to see if there is any milk residue left on it. Milk from a healthy cow will leave the strainer perfectly clean if you washed the cow. A cow can look fine and still have mastitis.

Mastitis usually affects only the quarter germs have entered, since the quarters are separate from one another. Mastitis can damage the milk-producing glands so much that the quarter can no longer make milk, becoming a "blind quarter." A bad infection may even cause the death of the cow.

You can test milk with a kit (like the California Mastitis Testing Kit) that measures the amount of somatic cells in the milk. Somatic cells are white blood cells and mammary cells damaged by the infection. The somatic cell count in the milk will be high as long as there is any infection in that quarter.

Ask a veterinarian or county Extension agent where to find a test kit. Complete instructions come with the kit. It contains a plastic paddle with four shallow cups for samples of milk from each quarter. A special reacting agent is put into each milk sample and stirred with the paddle, creating a change in the milk if somatic cells are present.

Your veterinarian has antibiotics for mastitis. A syringe with a long plastic tip is gently inserted into the teat so medication can be squirted into the quarter. Follow directions on the label for proper use and the number of days' treatment. A quarter should be thoroughly milked out before putting in the medication.

Milk from an infected cow should not be used for humans. Even after the infection is cleared up the milk can't be used until there is no more antibiotic residue. The medication label tells how long to wait. In the meantime, milk the cow regularly to hasten her recovery. In most cases, milk from a recovering cow can be used for feeding calves.

Milk Care

Take care to keep milk very clean. Use a clean bucket, clean strainer, and clean containers. You can purchase disposable strainer pads inexpensively through a dairy supply catalog.

Wash equipment and containers thoroughly after each use. Rinse all equipment in warm water to remove milk fat; cold water doesn't work as well. Then scrub with hot water and dishwashing soap, using a stiff plastic brush. The brush will clean the equipment much better than a dishcloth. Always use a plastic brush or scouring pad rather than metal. The metal can scratch the surface and leave tiny indentations where bacteria could cling. Rinse thoroughly in clean water.

You may prefer raw milk, although in some states it is illegal to sell raw milk, or you may wish to pasteurize the milk for your family's drinking supply to make it perfectly safe (see chapter 15).

The Lactating Cow

Proper nutrition determines how much milk a cow will give. She can't produce milk without the building blocks provided by adequate amounts of protein, calcium, and other nutrients. About two-thirds of the dairy cow's total nutrition should come from forages — hay, silage, or pasture. Make sure the quality is very good. Cows will eat more total volume of hay or silage if fed fresh feed several times a day. The more good forage a milk cow eats, the more milk she will give.

Silage is created by cutting the whole corn plant and storing the crop in a pit or upright silo where the plant material ferments, enabling it to keep without much loss of nutrients. Haylage is created by cutting legume or grass hay and storing it the same way. Both can be fed to dairy cows.

Cutting hay before it is overly mature and storing it as haylage or low-moisture silage has the advantage of needing less time to cure and reducing labor in harvesting because there is no baling. It also keeps more of its nutrient value due to less drying and leaf loss. It must be stored with proper moisture conditions, however. If it is too wet, it will ferment too much and the cows won't eat as much of it. If it gets too dry before storing, it will not ferment properly and may mold or heat too much.

If you live in a region where corn is grown, corn silage can be an excellent source of feed for dairy cattle because it contains about

50 percent grain (corn) as well as roughage. If harvested and stored properly, cows will eat a lot of this feed and merely need a protein and mineral supplement to balance their diet. For maximum yield, the corn should be harvested when cobs are mature but the plants are still green. If the plants are immature, the silage is usually too wet and also the field will have less yield per acre. If the plants become too dry before harvest, there is also less yield, and the silage is of poorer quality, with poor compaction, more mold, and lower palatability.

Although a dairy cow has a large rumen, a high-producing cow can't eat enough forage to supply her needs. For top production, she needs grain. Your county Extension agent or a local dairyman can advise you on a feeding program to create a well-balanced diet for your cows using locally grown feed, which is less expensive than feed transported a long distance. The amount of grain for each cow should be adjusted to fit her needs. She needs more during the peak of lactation when she is making the most milk, and less grain toward the end of that cycle.

Water is just as important as feed. A dairy cow needs a constant supply of clean water. Milk is made up mostly of water. A cow needs three to five gallons of drinking water for every gallon of milk she produces. This "water" includes the moisture in her feed. A cow eating hay will need more drinking water than a cow on lush green pasture that contains a lot of moisture.

Cows give less milk during cold weather if they don't drink as much as usual. Make sure their water supply never freezes up or gets too cold. Cows will give more milk in winter if the water is heated and kept above roughly 50 degrees.

Rebreeding the Dairy Cow

A cow should be rebred about three months after she calves so she will calve again next year at approximately the same time. Some cows come into heat less than a month after they calve, if they are well fed, but this is too soon to breed them. Allow her body to recover from calving; don't try to rebreed her until her calf is at least two months old.

Keep track of her heats when she starts cycling again, so she can be bred at the proper time. It may be several weeks before she starts having heat cycles, but you must watch. One clue, if you are milking her, is that she will have a temporary drop in milk production the day she is in heat.

If there is anything unusual about your cow's cycles, or if she has a discharge of pus from the vulva, have a veterinarian check her. If she has a uterine infection, it must be cleared up before she can have another calf; discover it early and treat it before she is bred.

After your cow has been bred, have your veterinarian perform a pregnancy check if she does not return to heat. It helps to know if she is pregnant and when she will calve next year so you can plan the proper time to dry her up before her next calving. If she is not pregnant, you must try to breed her again.

If you didn't have her tested for pregnancy, you may still be able to tell if she's pregnant. After the fifth month of gestation, the fetus will be large enough that you can feel it kicking. As you lean your head into the cow's right flank during milking, you may feel a bump on the head by a small foot!

Managing the Dry Cow

To be ready for her next calf and new milk production, the cow's body needs a rest from making milk. She should go dry for about two months before her next calving.

The actual length for the dry period can vary, depending on the cow's age and body condition. She needs at minimum 45-day rest to be able to produce a lot of milk during her next lactation and to make enough antibodies in her colostrum for the next calf. A six-week dry period may be adequate for the average, healthy cow, but a young cow, such as a first calver, and very high producing cows generally need eight weeks (56 to 60 days).

To help a cow dry up, reduce her feed, especially the grain. Eliminating grain helps her body adjust to not making milk, since she needs the extra calories for high milk production. Check her udder closely while she is drying up to make sure there's no heat or swelling. After the last milking, treat each quarter with an antibiotic recommended by your veterinarian to help prevent mastitis. If she does get mastitis, it must be treated.

When you dry up a cow, do it abruptly. Just stop milking her. It doesn't help to try to ease into it by partial milking. Mother Nature programmed the cow to stop producing milk when her udder is full and tight, as would happen under natural conditions if a calf dies. Pressure in the udder signals her body to stop making milk. Trying to help a cow dry up by milking a little out to relieve the pressure only prolongs the process and makes her more susceptible to udder problems and mastitis.

She will be uncomfortable for a while, but after a few days the pressure will ease, especially if you cut back on rich feeds like grain and alfalfa, which she has been utilizing to make milk. Her body gradually reabsorbs the milk that was left in her udder.

Pasture or hay should provide enough nutrition for her to go through the dry period without becoming fat. She will need grain only if she is thin and needs to gain weight before her next calf.

Raising Extra Calves on a Nurse Cow

If you don't need all the milk from a cow, she can raise extra calves. This can increase the income from a dairy cow and eliminate the chore of milking if you just want to raise calves. A good nurse cow can raise four to eight calves each year.

A nurse cow doesn't have to be a high-producing dairy cow. A good crossbred cow (half beef, half dairy) can do this, though she may not be able to feed as many calves. If the first calf you raised is a crossbred dairy-beef heifer, she could make a good family milk cow or a nurse cow.

Extra calves can be purchased cheaply at a dairy after they are born (see chapter 11) and raised with little effort, since the nurse cow feeds them. A mature dairy cow could raise eight big beef steers each year, or eight good dairy heifers. A nurse cow can raise two sets of calves

A nurse cow can usually raise three or four calves at once.

each year, if calves are fed hay and grain before weaning, so they can be weaned at 4 to 5 months of age. The cow will produce milk for 9 or 10 months before you dry her up for her next calf. You can wean the first set of calves halfway through her lactation and put four new ones on her. First-calf heifers may not be able to raise as many calves; her own and one extra may be enough for her to feed the first time. If you only have room for one or two cows, this can be a way to maximize what they produce. Selling eight calves a year from each nurse cow is easier than milking them and selling milk, and usually more profitable.

You can also compromise and enjoy the best of both situations by sharing the family dairy cow's milk with just one or two calves. If your cow gives more milk than your family can use, and you don't want to sell the extra, let her raise her own calf and an extra one while you take part of the milk. Let the calves nurse one side (each one on a quarter) while you milk the other, or milk your side first and then let the calves have the other side.

When raising calves on a nurse cow, you need pens for the calves next to the milking area. If raising dairy heifers, follow the management practices outlined in earlier chapters; they must be in individual pens to keep them from sucking on one another after each nursing.

You can buy extra calves just before the cow freshens and feed them on bottles until she calves, or buy them just after she calves.

Whenever a cow calves, keep and freeze a little colostrum to give to any future newborn calves you might buy that have not had adequate colostrum when you buy them.

Train the cow to accept "foster" calves by feeding her grain in her stanchion or tie-up spot and start the calves on her while she is eating. If she loves her own baby, she won't kick it, but she may try to kick off the extra calves. It takes time and patience to convince some cows to accept all the calves nursing at once. Supervise the nursing to make sure they don't bunt her udder or she doesn't kick them. If the cow kicks too much, she may discourage a timid calf. You might have to put hobbles on the cow before you let in the calves.

Always put the same calf on the same teat. If each calf has its accustomed place at the udder, it is not such a circus of confusion when you let the calves in with the cow. Be firm and make each calf learn its proper place.

Let calves nurse until they are done. You can tell the quarter is empty when the calf starts bunting the udder or tries to steal a teat from one of the other calves. Don't leave them on too long or they'll bunt in their greediness for more milk. Bunting calves can bruise the udder. You must be the referee. If a calf bunts a lot as he nurses, stand nearby with a stick and reprimand the calf with a sharp rap on the head each time he bunts, until he learns not to shove. The cow will kick at him if he bunts, unless she is hobbled, but she may hit the wrong calf.

The hardest part of the job is to make the calves leave the cow and go back to their pens after they have finished nursing. You may have to use a stick to encourage them to behave and go back where they belong. You can put halters on the calves if you wish, before you let them out of their pens. Leading lessons will make them more manageable, as when taking them away from the cow and back to their pens. Double the lead rope over the calf's back while he is nursing the cow, so it is off the ground and won't be stepped on.

CHAPTER FIFTEEN

A Dairy of Your Own

I F YOU WISH TO SELL MILK, you must take good care of the cattle and practice extreme cleanliness to make sure milk is always clean and safe to drink. In some geographic areas, small dairies can be profitable, but you first need to do your homework to see if this will work for you.

Starting Your Dairy Business

Start by getting advice from local sources. If you want to start a dairy, visit your county soil conservation or natural resources council and county extension office to find out how to comply with local laws and regulations. If you plan to sell milk, also check state and federal regulations. If you own or rent land, you will be doing some land and crop management. You can get advice from local agencies for putting in crops and deciding which crops work best in your region, climate, and soil; as well as advice on building facilities and irrigating if you live in a dry climate. Learn about regulations regarding manure management and other factors involved with farming.

It helps to have a network of sources for advice and help. These sources may include established farmers in your area, including your parents or relatives. An organization like a dairy breed association, 4-H council, grazers' association, or any other group of people you can rely on for information and advice can be very helpful. Family members who are willing to loan you money to get started, or to bail you out in hard times until you get on your feet, can certainly be a plus.

If you have no previous farming background, it's also very beneficial to acquire hands-on experience before you plunge into this. Check with your county Extension agent to see if there are any farmers in your area who need part-time help. Some farmers are better to work for than others, and your county Extension office can probably recommend one. You might work evenings and weekends and still keep your other job until you know if you really want to farm full time. Real-life experience is invaluable; these lessons often sink in more than when you merely read or have someone tell you about it.

If you want to start a dairy, you need to be in a region where dairying is profitable. Some outlying rural regions are too far from markets to make it a viable enterprise; transportation costs for hauling milk to a processor or sending your milk products to market may be too high. You also must be close to your feed supply or live in an area where you can grow your own feed. If you have to haul grain or dairy-quality hay a long distance, this may make the milk too costly to produce.

Your Business Plan

To be successful at farming, you need good business management skills. Before you start, you need good financial projections. After you start, you must keep good records and learn some basic accounting. You might be able to produce a lot of milk per cow by taking excellent care of your cattle and feeding them well, but if it costs too much to get the extra pounds of milk, you won't make money. If you spend more (for feed) to get the extra pounds of milk than what the milk is worth, you'll lose money. It may be worth it when milk prices are high, but if milk prices are low it won't pencil out.

You must be able to continually assess what you are doing and have clear objectives. If you want to start a dairy, put together a solid business plan — what your goals are, how you plan to accomplish them, and what the projected costs and capital investment will be. Find an advisor to help you put this together; you are not just producing milk for the fun of it. Even though your main objective might be to live on a farm, you don't want to go broke doing it. You may need to find ways to diversify your business, such as raising dairy heifers for market or selling compost (see box above). The small dairy needs to look at every one of its resources to see how these can be turned into money.

The financial challenges can make or break a farming effort, and you need to know whether it will work. Having some equity is helpful but not required. A young person may start by renting, build up an equity in cattle and equipment, and then borrow to start buying land. In contrast, a person who has had a 20-year career doing something else may have made enough money to get started without borrowing.

The Basics

If you want to keep dairy cows, you'll need some type of shelter from bad weather, so a barn is a necessity. Depending on your location and your goals, you may also want to raise your cows on pasture for at least part of the year, in which case you will need to know what type of fencing to use.

A Barn

You need a clean place to milk the cows in bad weather, so a barn of some kind is essential. It doesn't have to be elaborate. An old barn or building can often be converted into a small dairy barn and you can install milking equipment if you have more than one or two cows. A concrete floor is easiest to keep clean, but a wooden floor made of well-treated thick planks that won't rot will do. The barn should have windows and good ventilation. A tightly closed-up barn is unhealthy, holding in moisture and ammonia fumes from manure and urine. These can be damaging to the lungs and the cattle will be more susceptible to pneumonia.

Box stalls for cows in the barn for calving or housing a sick cow should be at least 10 feet by 10 feet. For milking and other purposes, such as administering treatment or any instance in which you need to restrain the cow, smaller tie stalls or stanchion stalls work. It takes much less bedding for a tie stall or stanchion stall than a box stall, since cows are confined to one spot and manure will all be deposited at the rear. Tie stalls or stanchion stalls don't need walls. Simple rail dividers between cows work fine; you can duck under it if a cow becomes fractious.

The tie stall or stanchion area should have a slight slope to the rear so liquid from manure and urine will stay behind the cow and not soil her udder if she lies down. Many dairy milking areas or tie stalls have a depression or gutter along the back, behind where the cows stand. This creates a place for fluid to run and makes it easier to clean up behind the cows.

If you use silage, you can often find a used upright silo to buy. The dairy industry is now using flat storage more commonly, but flat storage is not feasible for a small herd. It takes up more space and you are not using enough volume daily to keep the exposed portion fresh.

If you keep hay or grain in the barn, use a safe, separate area from where the cows are. You won't want loose cows in the barn getting into the feed. Store grain in mouse-proof containers, not in bags on the floor. Large garbage cans with lids will work. If you have several cows, you may want to buy grain in bulk and use a storage bin.

With only one or two cows, you can milk by hand, take care of the milk in your kitchen or a back porch utility area, and store it in a springhouse or refrigerator. If you have a number of cows, and milk is being picked up or taken to a creamery or some other processing place, you'll want a small bulk tank in the barn for cooling and storing it. You can buy a used tank from a dairy that's getting a larger or newer one. Talk to a dairyman about what type and size tank might work.

You'll also need milking equipment if you have more than a few cows. A beginning dairy can usually find what is needed by checking the used equipment market and may be able to get by with bucket milkers. The field person at the processing plant can advise you regarding the necessary equipment and how to set it up. The plant probably has some producers who are selling functional older equipment.

Pasture

Having your cows graze in a pasture part of the year saves a lot of purchased feed. Heifers raised on pasture are generally healthier and more athletic than confined, grain-fed heifers. Recent studies have shown the milk from grass-fed dairy cows contains more of the healthy omega-3 fatty acids and less of the harmful omega-6 (see chapter 7). The milk also contains five times more CLA (conjugated linoleic acid) than milk from cows on a grain diet.

Many small (and some large) dairies have switched from a high-grain diet to high-forage diet. This works best if you live in a region (such as the Southeast) with year-round grass. In northern areas, or some of the drier western states, pastures must be supplemented when they are not growing. Only a little more than 10 percent of dairy cows in the United States are grazed.

The investment required for a pasture dairy is usually less than for a confinement operation (less equipment and buildings), and a grazing system can be very feasible on a family-size farm. The lower capital investment may be attractive when a person is trying to get started. A family can potentially make a living on a small grazing farm that can be operated by family labor.

If you have a winter season when grass is poor or unavailable, you can still have a predominantly grazing dairy by feeding the cows intensively for only part of each year. You may also choose to do seasonal calving, with all calves born in spring, and drying off all cows during winter, so you are not milking in winter. Then you won't have to feed cows so heavily during this season when they cannot graze. Even though research has shown that seasonal calving and milking generally results in lower annual profitability than year-round milking, it's an option for people who want less labor-intensive winters with a grass-based dairy.

Fencing

You will need different types of fences for different purposes. Moveable electric fence works well for rotational grazing; only boundary fences must be sturdy, solid fencing so the cows don't go trooping off to the neighbor's place, out on a highway, or into your garden.

Grassy areas that are not under fence, such as along a roadway, can be utilized with a gentle milk cow that is trained to tie. She can be tethered to a stake in those areas and moved periodically throughout the day. Make sure a staked cow has shade during the heat of the day and periodic access to water. Don't use a really long rope; usually 20 feet or less is best. Move her often. Cows tend to eat at the end of the rope and tromp out the grass toward the center. A long rope and a large circle only encourages her to waste more grass.

It's better to use a stake than to tie a cow to a tree. She will tend to wrap herself around a tree. Drive a piece of iron into the ground, deep enough to hold, with enough above ground that she won't lie on it or step on it and injure a foot.

The best use of a good pasture is to graze it in sections. This gives you the most feed, with less wasted. For dairy cows, often the easiest way to do this is with an electric fence, moving the fence every few days to graze the pasture in sections or strips. Cows continually have good, fresh grass, and the segments behind them can grow back. See chapter 1 for more information on fences.

Cow Care

Many things about cattle care have been covered in earlier chapters, but here are a few more tips on how to keep your dairy cows healthy and happy.

Milking Schedule

Some high-producing cows are milked three times a day, but the average dairy cow does fine if milked twice a day on a regular schedule. The schedule is very important — not so much the specific hours, but each milking should be about 12 hours from the last one. You can milk at 6 A.M. and 6 P.M., or 10 A.M. and 10 P.M., or whatever time suits you best as long as you adhere to the same schedule. (Studies have shown, however, that cows do not like to be milked before dawn, especially in the hours between 3 A.M. and 6 A.M. This is when they generally sleep, and they do not want to get up.)

Late milking causes a tight udder, which is nature's signal to halt milk production; this may cause a drop in milk. The cow may develop mastitis from bruising. A regular schedule avoids these problems.

A few small dairies have tried once-a-day milking, and it seems to work. If your time is limited, this can be a way to still have a dairy. The cows give less milk but also eat less due to less demand on the body, so more cows can be maintained on the same amount of feed. The cows tend to stay in better body condition: They do not become thin, and they breed back quicker.

Consider once-a-day milking if you have another job that interferes with a twice-a-day schedule; if you live in a hot or cold climate that is stressful to the cattle; if cattle have a long walk to the barn from steep pastures; or if your pastures are not dairy-quality but you still want to have a grass-based dairy.

Preventive Care

Clean housing and bedding are good prevention against dirty udders. Manure on udders can lead to mastitis if germs enter the teat canal. Older cows may have saggy udders or long teats, which are more easily contaminated or injured by other cattle stepping on their teats when they are lying down. To avoid injury, make sure cattle are never crowded. Have each cow in a separate stall or stanchion when in the barn.

Put a magnet into each heifer when she is a yearling to prevent "hardware disease." One magnet per animal is adequate. If you accidentally give a cow another magnet, the magnets will attract one another and cancel each other out if they happen to end up side-by-side. When buying a cow, ask if she already has a magnet before giving her one. Another way to check is to put a compass next to the cow's abdomen. If the needle moves, she has a magnet.

Cow's hooves grow continuously just like horse's hooves or your own fingernails and toenails. Cattle walking on dry or stony ground wear their feet down about the same rate they grow, but cattle on soft pastures may have problems with their feet growing too long. If toes grow too long, they curl forward and the cow's weight is all on the heel. This puts too much stress on leg structures, and she may go lame. Her base of support has grown out from under her; eventually she'll be walking on her dewclaws and the back of her pasterns.

Cows' hooves can be periodically trimmed, if needed, before they become this bad. If a cow is gentle, she will usually let you trim her feet with hoof nippers. If she won't let you pick up her feet, you can often trim the long toes while she's standing. If you don't want to tackle this job, ask your veterinarian to help. Some cows never need to be trimmed; others need a trim once or twice a year, depending on their rate of hoof growth and their environment.

Milk Production

A cow gives the most milk at the peak of her lactation — a month or two after calving. After the initial udder swelling disappears, her milk production increases dramatically during the first three weeks after she calves, reaching a peak between five and eight weeks after calving.

How long she stays at peak production depends on her genetic potential and how well she is fed. A good dairy cow with adequate high-quality feed generally maintains a high level of milk production for another four to six weeks before she begins to drop. Cows on lower quality feeds that are pastured rather than fed concentrates often reach their peak a little earlier, since their top production will be lower and they won't maintain it as long. They also drop off more rapidly after passing their peak.

High-producing cows in commercial dairies rarely last more than three or four lactations. They are fed for maximum milk production and many of them "break down" or burn out after a few years of milking. By contrast, the same type of cow in a small dairy where they are not pushed so hard may milk for 10 or more years. A key factor in how long a good dairy cow will last is her udder construction. If she has a strong, well-attached udder that does not break down, she may continue to milk for many years.

A cow's annual production of milk is highest when she is young, but if she is well fed and well cared for she may continue to milk adequately for many years. A cow that is not "burned out" by being pushed too hard in early years will continue to milk well until middle age. Her annual production may increase as much as 20 percent through her first two lactations, and more slowly through her sixth or seventh lactations (at age eight or nine), then slowly decrease.

One of the first signs of illness in a cow may be a drop in her usual milk production. A cow in heat may drop dramatically for one milking. If a cow does not bounce back at her next milking, and you are sure that she was not in heat, be suspicious that something is seriously wrong with her. Take her temperature and then have a veterinarian check her.

Lack of water will decrease milk flow. If the water supply is low or she drinks less for some reason, this is reflected by a drop in milk. Upsetting her regular milking schedule can also cause a drop in production. Anything that upsets her may influence the amount of milk she gives. If she gets nervous in the barn because of noises she's not used to, or strange people, she may give less milk.

Fat content in milk varies from cow to cow. Butterfat content (the amount of fat droplets and how large they are) is often directly related to how much milk a cow gives. Holsteins give the most milk and their milk is lowest in butterfat. Jerseys don't give nearly as much milk, but it is very high in butterfat. Each individual cow has her own capability and it is also influenced by the feeds she eats.

Cows often have a higher butterfat content in winter than in summer just because they give slightly less milk in winter. A seasonal change in feeds, colder weather, and reduced water intake can account for less milk in cold weather.

A cow has relatively high butterfat content right after calving; it decreases as overall milk volume increases. After milk production peaks and begins to drop off, butterfat content tends to rise a little again. Mature cows of the same breed have higher butterfat content than two-year-olds in their first lactation.

The richest part of any cow's milk is in the "strippings" that are milked out last. The first few squirts from her udder are very low in fat, but the last squirts are quite rich. A cow whose milk tests at about 5 percent butterfat will have 10 percent fat or higher in the final strippings.

The color of milk is not so much an indication of butterfat as it is carotene content. The white milk and cream of Holsteins, Brown Swiss, Ayrshires, and Shorthorns is due to converting most of the carotene they eat into colorless vitamin A. The milk and cream (and also the body fat) of Jerseys and Guernseys is more yellowish because of carotene from the feed.

Flavors in Milk

You don't want "off" flavors or odors in the milk you sell. The best way to prevent bad flavors in milk is to keep the milk perfectly clean and to make sure the cows do not eat plants that might create unwanted flavors.

The worst thing milk can taste like is manure. This can happen if you fail to have the cow's udder perfectly clean when you milk, or have dirty hands. If a cow is wet from rain, a few drops of water falling from her dirty flanks into the milk bucket is all it takes. Keep your cow, hands, the barn, and milking equipment clean.

Too much manure in the barn, or musty or moldy smells, can be picked up by warm milk. If milk has a "barn" taste, either your barn is dirty or warm milk is standing too long in the barn before it is chilled.

Some plants eaten by the cow can make milk taste bitter or strong. Avoid wild onions, china lettuce, horseradish, leak, garlic, some types of silage, cabbage, turnips, and ragweed. Remove offending weeds in your pastures. Make sure hay does not contain weeds. Some of the plants that cause problems are most obnoxious if a cow eats them within a few hours of being milked.

If the cow is heavily pregnant and you are still milking her, the milk may begin to taste a little salty or strange. A cow due to calve in eight weeks or less may drop off noticeably in production and milk will taste different. It is time to dry her up and let her prepare for her next lactation. A cow with mastitis also may have milk that tastes a little different or salty.

Product Safety

Milk from healthy cows is safe food, as long as it is kept clean and has no chance of becoming contaminated. It should be strained and cooled as soon as possible after milking.

Milk buckets and all holding containers and strainers should be perfectly clean. Scrub and rinse them after every use. Rinse with boiling water, if possible. Otherwise, use very hot water, and hang the bucket or strainer to dry in sunshine.

The sooner milk is cooled after it comes from the cow, the less chance for it to pick up off-flavor odors from the barn, and less chance for bacterial contamination. Milk keeps best at about 40°F. The quicker you can chill it down to that temperature, the better.

If you don't have room in your refrigerator for all the milk, you can use a springhouse or well if the water is cold enough — preferably less than 50°F — but milk won't keep as long. You may want to invest in another refrigerator just for the milk.

One way to help take the load off the refrigerator is to partially chill the milk first. Set the containers in cold water for a short while before putting them into the refrigerator. Change the water a couple of times in a sink, refilling with cold water after the original water becomes lukewarm from drawing warmth from the milk. This method can chill milk much more quickly than just putting it into the refrigerator when it is warm. Don't cover raw milk until it has chilled. This allows odors to escape.

Raw or Pasteurized?

In many states it is illegal to sell raw milk. You can drink raw milk from your own cow, but you cannot sell it. Check with your state authorities on the laws governing raw milk sales. Proper pasteurization of milk makes it safer to drink, if there is any risk of contamination. Heating milk for this germ-killing process will not kill all the organisms that might be in it, but it usually kills the ones that commonly affect humans and the ones that cause milk to spoil. Properly pasteurized milk will often keep longer without souring than raw milk.

When you sell milk, you must follow standards set by the department of agriculture in your state. A few dairies do on-farm processing (if the milk must be pasteurized), but most of them sell to another processor. You also must meet the standards set by the plant or co-op that buys the milk. The plant salesman or field person is usually willing to spend time advising new producers.

Milk must be periodically tested and in compliance with all regulations. Milk testing is a state (rather than a federal) responsibility. The Pasteurized Milk Ordinance is a document issued by the Food and Drug Administration, incorporated by the states into their laws or regulations. In Massachusetts, for instance, there is a state law that spells out which agency has the authority to regulate certain aspects of the dairy business. The Department of Agricultural Resources in Massachusetts regulates milk production from the farm up to the point when it enters the processor. After that it is under the jurisdiction of the state department of health.

In most states, milk can be sold for public consumption only if it is pasteurized or sent to a processing plant where it will be pasteurized for drinking or used for making cheese and other products. To sell milk from your own dairy to the public, it usually must be Grade A (pasteurized) milk. In many states, cheese factories will buy Grade B (unpasteurized) milk, and there are many differences in the requirements for the Grade A versus Grade B dairy (such as water quality for the cows, housing restrictions, and design of the milk house and barn). You need to find out what is required in your own state, since this is not a federal issue.

In states where it is allowed, raw milk sometimes brings a premium because it is considered a health food ("natural"). Raw milk, for instance, is full of enzymes and vitamins that are destroyed by pasteurization (for more information, see Helpful Sources, page 253). If you want to sell raw milk or cream, check with your state authorities about the regulations and testing requirements, and talk with someone who markets milk this way to get tips on how to properly care for the milk.

If your state allows the sale of raw milk at the farm for retail purposes, it must be tested for quality (checking bacterial count, antibiotic residues, and so on) and the regulations are quite stringent. You must have signs on the premises and labeling on the product to adequately inform customers that the product is not pasteurized. In Massachusetts, for instance, anyone producing milk for commercial purposes needs a certificate from the State Department of Agriculture; if dairies want to sell milk retail they must inform the department so that testing can be done.

If milk or milk products are sold at the farm, samples are taken there. If the milk is marketed through a co-op or a processing plant for making cheese or some other product, samples are taken from the truck hauling the milk to the plant. The hauler also takes regular samples, since payment to individual farmers depends on these samplings. From these samples, they measure such things as quantity (pounds of milk being picked up), butterfat content, true protein and other solids, and make sure there are no antibiotic residues. A co-op may offer premiums for quality characteristics as well. If the milk has a low somatic cell count, you get a higher price for it.

If you want to use or sell pasteurized milk, there are several ways to pasteurize it. Milk can be heated to a full rolling boil for a couple of seconds, but this leaves it with a cooked odor and taste (scalded). You can heat it to 170°F and keep it at that heat for 15 minutes, or you can heat it to 150°F and leave it at that temperature for 30 minutes. The latter does not affect the flavor as much, especially if you heat it in a water bath instead of over direct heat.

Home-pasteurizing units consisting of a metal container with a heating element in the bottom can be purchased through a dairy supply catalog. Fill the container with water to a certain level, set a gallon of milk in it (in a covered pail), surrounded with water, and plug it in. The water heats to the desired temperature for the proper length of time, and a buzzer sounds to tell you it's done.

An electric pasteurizing unit with timer is the easiest way, but you can also pasteurize milk with quart jars in a home canner water bath. Pour the strained milk into sterilized jars, filling them to within an inch of the top. The uncovered jars of milk are placed on a rack in the canner. The canner is filled with warm water to just above the milk line in the jars. A dairy thermometer is suspended halfway down the center jar of milk, so you can determine when the milk is at proper heat. Heat the canner on the stove until milk reaches 150°F for 30 minutes or 170°F for 15 minutes.

After the appropriate length of time, lift the rack of jars into a sink of lukewarm water. You don't want water too cold, or hot jars may break. Change the water in the sink several times, putting in colder water as milk begins to cool. As soon as the milk has dropped to room temperature, put clean lids on the jars and put them in the refrigerator. If you want to chill them more first, you can put colder water or ice cubes in the sink water.

Raw milk or cream will keep at least four or five days if it is clean, chilled quickly, and kept refrigerated. Pasteurized milk or cream should keep for more than a week. If it sours more quickly than it should, this may be a sign your equipment isn't clean. Another cause for milk to spoil sooner than usual is if a cow has an infected quarter (mastitis). Germs in the infected milk will cause it to go sour.

Separate the cream from the milk by letting the cream rise on its own, which takes several hours, or by using a mechanical separator. Separators work best if the milk is still warm, before the cream has started to rise. When taking the cream off a lot of milk, you might want to use a separator; otherwise it's not worth the effort of cleaning it. It has to be taken apart and scrubbed and sterilized after each use.

Direct Marketing of Milk or Milk Products

Some small dairies have been very successful in selling their own milk or milk products in a niche market, but this adds another dimension to your farming operation that isn't always easy to accomplish. There is not only the challenge of operating the farm and producing the product; you also must be able to market the product. All too often people have a dream of retiring from the city to the quiet countryside to produce a wonderful product, thinking customers will beat a path to their door, but many have tried this and been disappointed. The type of individual who does well in farming is often an introvert and may not do as well as a salesperson. You must be realistic about your market and be located where people will find you.

If you are close enough to large population centers, you may be able to market your products through your own farm store. Some New England states, for instance, have a large number of small dairies that do direct marketing. You can bottle your milk yourself or make a cheese, yogurt, or ice cream product. There are some very successful cheesemakers in New England and Wisconsin, for instance, who use their own milk.

Whether you want to sell raw milk, cheese, or ice cream, or become a producer/dealer, you need to have a good marketing plan and be sure of your customer base. You can't put up a store in the middle of nowhere and expect people to come. Define your market before you start. It has been estimated that less than 20 percent of shoppers will take the extra time to go to a farm-based store; most consumers want the convenience of buying everything at one big store. You need a large population base to make a farm store work. For those shoppers who do make the extra effort to come to your farm store, they are buying more than just your product: They are also purchasing the farm experience. Developing relationships with your customers is key to making your farm store successful. Make your displays attractive, be welcoming to customers, and even plan regular events that will draw more people to your farm.

Combining Wholesale and Retail

Even if you depend on sales at your farm for most of your income, you need a plan for what to do with any milk produced in excess of what you can sell retail. That milk can be sold through some kind of marketing arrangement with a co-op or a milk hauler who takes it to a processing plant.

On the other hand, a small dairy may not want to depend entirely on wholesale business. In many cases the successful ones have something they produce and package themselves and sell retail at their farm. If you depend on a chain of health food stores to market your product and you suddenly receive a notice they are ceasing operations, you have no market.

Do your homework and market research, identify what could potentially be your customer base, and have a realistic approach for going after it. The location of your dairy, farm store, or home delivery milk business, is very important. There are niche markets that a small dairy can fill, while building some level of wholesale business to make a living. If you identify your market, discover the potential is there, and go to a bank with a solid business plan, you won't have much problem putting together the captial to build some type of small processing facility on your family farm to produce the product you want.

How Many Cows?

To make a living with a small dairy, you usually need at least 40 cows, depending upon your situation, location, and the market for your milk. In regions where there are many small farms and close access to markets, small dairies can work. In Wisconsin and some New England states, for instance, the average-sized dairy herd is only 70 cows. A few farmers can make it with only 25 cows, but they live very close to the land and don't spend anything on themselves. Most people who try it on this small scale, however, are not as spartan as they imagined themselves to be and discover they need more cows to make an adequate living. Depending upon where you live, however, and the prices for certain products in health food or gourmet markets in your region, a person may be able to make an adequate living selling raw milk, gourmet cheeses, or some other desirable product from just a few cows.

You don't need to start with "optimal" size at first, however. Some people keep another job for a couple of years, raising heifers on rented land, building up equity and experience before plunging into their own dairy. Forty to 60 cows is about the range that one family can handle. Grazing dairies can sometimes manage more cows than a confinement operation. If cows are well-managed and you haven't paid high prices for cattle and land (and have made some good business decisions), it is possible to make a living from a 40-cow dairy. It is generally a mistake to start with just 20 cows and no other income, thinking you can do that for a few years before increasing your herd.

NICHE MARKETS

Only a very small percent of the population is interested in buying raw milk. This is an extremely small market. You might sell milk from two or three cows this way, but to do it on a large scale you'd have to live near a highly-populated area to draw enough interest for major sales. Likewise, there are a few people who are interested in buying unhomogenized milk (in which the cream rises so you can take it off the top), but most small dairies do better if they go after a larger customer base.

If the farm is going to take most of your effort, it should reward you for that effort. In many situations one member of the family works off the farm and that job is the sole support of the farm; the farm may not be breaking even by itself. Sometimes one spouse works off the farm and the other is farming full-time and together they do well, but you have to figure out what you want to accomplish. Many beginning farmers find it works best to buy some heifers and work into the dairy business gradually.

Will Dairying Work for Me?

If you decide to go into dairying, you need to do your homework on the financial part. Closely scrutinize your goals regarding herd size and your own commitment. Even though a small family dairy may be what you want, this takes a major commitment, often with longer hours than a large dairy in which more people are involved. For a small herd, one advantage is that you can often use existing facilities on a purchased or rented farm, without making a lot of changes, but this type of operation is also very labor intensive. What may sound like fun up front requires a major time investment — seven days a week, with long hours.

If you have a family, you need to consider what the family wants to do. It's very important for all of you to have the same commitment; farming is very different from other occupational choices. In most jobs, the husband or wife goes off and works, spending eight hours away from home, but this doesn't necessarily change what the children do. When you move your family to the farm, this makes a major change in where they live and what their lifestyle will be.

Your family is involved, to some small degree or a lot, in your occupation. This is one reason many people move to a farm — to raise the kids in the country rather than the city. But you need to be conscious of this when making this decision. It will involve and impact all of you, on a day-to-day basis — not just the fact that you are earning less money, but also that you are living in a place with chores to do every day, all year long. If your kids enjoy this, it can be a wonderful experience. If they don't, it may not work. The decision to farm must be embraced by the whole family.

One farmer very aptly described farm life as a book with a wonderful cover. A lot of people look at the cover and think they would really like it. But when they get into about the second chapter, they find it is much more challenging than they expected. So to increase your chances for success, do your homework first, know what you are getting into, and make the plunge only after careful consideration. If you are interested in taking a university course, see the Helpful Sources section on page 253 for more information.

Gallery of Breeds

There are dozens of beef breeds and a half dozen major dairy breeds. Each breed has unique traits that set it apart from the others. Some breeds are very old (the Chianina, from Italy, goes back 2,000 years or more, to the time of the Roman Empire), while others have been created in the past several decades by selectively combining the genetics of older breeds. Today the many beef breeds have some differences in size, carcass traits (lean or fat), weather tolerance, hair length and color, markings, and so on. Most are horned and some are polled. Some of the horned breeds have had Angus genetics infused into some of the herds, and then have been selectively bred, so the offspring are polled and black: two traits that are popular with today's stockman.

Beef breeds are stockier and more heavy-muscled than dairy breeds. The latter have been selectively bred for milking ability rather than meat and are finer-boned and more feminine, with larger udders. Beef cattle were originally bred for size and strength to be draft animals (oxen) to pull carts, wagons, and plows. When cattle were no longer needed as much for draft purposes, these large heavy-muscled animals were selectively bred to create better beef.

The farmers who raised certain types of cattle eventually created "breeds" with registries, as well as rules for members to follow when selecting and breeding them. These rules helped insure uniformity

and embodiment of desired traits. Many breeds started as "dual purpose" cattle, used for meat and milk (Brown Swiss, Shorthorn, and Simmental are good examples) and were later split into separate registries as either beef or dairy.

Even though some breeds are similar in color, they are not the same in other traits. If you are familiar with breed type, you can differentiate between a Red Angus, Limousin, Saler, Terentaise, or Gelbvieh, for instance, even though all of them are "red." Differences in body build, frame size, and "bone" (large bones or small) are more important than color in distinguishing the individuals of different breeds. Each breed has its ideal type that encompasses the traits stockmen select for when producing that breed.

The following photo gallery shows most of the major beef and dairy breeds, but there are many more beef breeds and several other dairy breeds. Some of the beef breeds not shown here are Barzona, Beefalo, Braford, Brangus, Braunvieh, Charbray, Chianina, Corriente, Devon, Marchigiana, Piedmontese, Romagnola, Santa Gertrudis, Senepol, Scotch Highland, Watusi, Wagyu, Welch Black, and White Park. Some of the other dairy and dual-purpose breeds are the Dutch Belted, Kerry, Milking Devon, Normande, Norwegian Red, Red and White, and Red Poll.

American Shorthorn

Cows 1,500 LBS / BULLS 2,000 LBS

Courtesy of American Shorthorn Association

ORIGIN: *northern England.* A dual-purpose breed (also called Durham); it is red, roan, white or spotted, and the calves are small at birth but fast growing.

Angus

Cows 1,100 LBS / BULLS 1,800 LBS

Courtesy of American Angus Association

ORIGIN: *northeastern Scotland.* Polled, black, and noted for high meat-to-bone ratio, meat quality, and fast finishing.

Beefmaster

COWS 1,300 LBS / BULLS 2,000 LBS

Photograph by Watt Casey, Jr.

ORIGIN: *southwestern United States*. Mix of Brahman, Shorthorn, and Hereford, creating a heat-tolerant breed with good meat production. Various colors.

Blonde d'Aquitaine

COWS 1,600 LBS / BULLS 2,500 LBS

Courtesy of American Blonde d'Aquitaine Association

ORIGIN: *southwestern France*. Light tan and similar to Charolais in size and physical appearance; noted for fast growth, carcass quality, calving ease, and fertility.

Brahman

Cows 1,200 lbs / Bulls 1,900 lbs

ORIGIN: *southwestern United States*. Developed from several strains of imported Zebu cattle. Calves are small at birth but grow fast.

Charolais

Cows 1,600 lbs / Bulls 2,500 lbs

ORIGIN: *central France*. White/cream with thick muscles, good feed efficiency, and heavy weaning weights.

Chiangus

Cows 1,500 lbs / Bulls 2,800 lbs

Photograph by Alan Sears

ORIGIN: *California, 1970s.* A composite "breed" of large, black cattle produced by crossing black Angus with Chianina (which come from Italy and are the largest cattle in the world).

Dexter

Cows 700 lbs / Bulls 900 lbs

Photograph by John Potter, Spruce Grove Farm

ORIGIN: *Ireland, 200 years ago.* These are the smallest cattle; docile and easy to handle, they produce rich milk and high-quality meat.

Galloway

COWS 1,150 LBS / BULLS 1,900 LBS

© Patricia Anderson Pruitt

ORIGIN: *Scotland.* Black, red, brown, and white (sometimes belted), these polled cattle have a heavy winter coat, are long-lived (15 to 20 years), and have an easy time calving.

Gelibvieh

COWS 1,400 LBS / BULLS 2,200 LBS

Courtesy of American Gelbvieh Association

ORIGIN: *Austria and West Germany.* Dual-purpose cattle, tan or gold, fast-growing, and maturing more quickly than most continental breeds.

Hereford

Cows 1,200 lbs / Bulls 2,000 lbs

Courtesy of American Hereford Association

ORIGIN: *England*. Draft animals with large frames and heavy bones; red body with white face, feet, belly, and crest.

Limousin

Cows 1,350 lbs / Bulls 2,400 lbs

Courtesy of Limousin World

ORIGIN: *western France, several thousand years ago*. Red or gold and well-muscled, these cattle have good marbling. They are finer-boned than Charolais but grow as rapidly.

243

Maine Anjou

COWS 1,700 LBS / BULLS 2,650 LBS

Courtesy of American Maine-Anjou Association

ORIGIN: *northwestern France.* This breed is the result of crossing French Mancelles (red and white draft breed) with Shorthorn. The mix is hardy and fast-growing.

Pinzgauer

COWS 1,350 LBS / BULLS 2,150 LBS

Photograph by Pamela Shay-Bryant

ORIGIN: *Austria (1600s).* Moderate-sized dual-purpose breed with unusual markings (dark body with white topline and "stripe" from front belly to back leg).

Polled Hereford

Cows 1,200 lbs / Bulls 2,000 lbs

Photograph by Christy Collins

Origin: *Iowa, 1901.* Foundation stock were mutant Herefords without horns.

Red Angus

Cows 1,100 lbs / Bulls 1,800 lbs

Courtesy of Red Angus Association of America

Origin: *United States, 1950s.* Developed from Angus that doubled the recessive red gene. A new breed was created with strict selection and breed standards to produce uniformity.

Salers

Cows 1,450 lbs / Bulls 2,450 lbs

ORIGIN: *south central France*. Popular for crossbreeding, these dark red cattle have good milking ability, fertility, calving ease, and hardiness.

Simmental

Cows 1,400 lbs / Bulls 2,500 lbs

ORIGIN: *Switzerland*. These dual-purpose cattle are noted for rapid growth and good milk production; the North American version is a beef breed.

Tarentaise

COWS 1,100 LBS / BULLS 1,900 LBS

ORIGIN: *French Alps.* Related to Brown Swiss, these moderate-sized dual-purpose cattle are cherry red with dark ears, nose, and feet, noted for fertility and early maturity.

Texas Longhorn

COWS 900 LBS / BULLS 1,450 LBS

ORIGIN: *feral Spanish cattle in the American southwest.* Hardy, long-lived, and noted for calving ease, but have less muscling than other beef breeds.

Ayrshire

Cows 1,200 lbs / Bulls 1,800 lbs

© Agri-Graphics, Ltd.

Origin: *Scotland*. This moderate-sized breed has red-and-white spots with jagged edges and is hardy and long-lived, with good udders. Gives rich white milk.

Brown Swiss

Cows 1,400 lbs / Bulls 1,900 lbs

© Agri-Graphics, Ltd.

Origin: *Switzerland*. Large and sturdy, long-lived, light to dark brown or gray in color, giving milk with high butterfat content.

Guernsey

Cows 1,100 lbs / Bulls 1,700 lbs

© Agri-Graphics, Ltd.

ORIGIN: *Island off the coast of France, 1,000 years ago.* Tan color with white markings (and yellow skin). Cows are medium-sized and give yellow milk rich in butterfat.

Holstein

Cows 1,500 lbs / Bulls 2,000 lbs

© Agri-Graphics, Ltd.

ORIGIN: *Netherlands, 2,000 years ago.* Large black and white cattle (or red and white), giving large volumes of milk that are low in fat.

Jersey

Cows 900 lbs / Bulls 1,500 lbs

ORIGIN: *Jersey Island in the English Channel.* Small cattle, fawn/cream, brown, moose gray, or black in color (tail, muzzle, and tongue are usually black). Milk is rich in butterfat.

Milking Shorthorn

Cows 1,500 lbs / Bulls 2,000 lbs

ORIGIN: *northeastern England.* Large, hardy cattle. Red, red and white, white, or roan in color. Milk is richer than Holstein but not as rich as Jersey or Guernsey.

Bibliography

BOOKS

The Family Cow, by Dirk van Loon. Storey Publishing, 1976.

Storey's Guide to Raising Beef Cattle, by Heather Smith Thomas. Storey Publishing, 1998.

Livestock Breeds of the United States. Agricultural Resources and Communications, Inc., 1995.

Veterinary Medicine, 9th edition, by Radostitis, Gay, Blood, and Hinchcliff. W.B. Saunders, 2000.

Your Calf, by Heather Smith Thomas. Storey Publishing, 1997.

INTERVIEWS

Glen Elzinga, rancher raising natural beef, Salmon, ID

Bill Gillmeister, MA Department of Agriculture

Eric Hart, manager, meat department, Good Food Store, Missoula, MT

Connie Hatfield, OR Country Beef, Brothers, OR

Tom Kriegl, Center for Dairy Profitability, University of WI

CJ Mucklow, Routt County Extension Agent, CO State University

Roger Palmer, Dairy Extension Specialist, University of WI

Warren Shaw, Shaw's Dairy Farm, MA

4-H Beef Project
PNW 448 — Pacific Northwest
Extension Publication
OR State University

4-H Dairy Cows and Management
North Central Regional
Extension Publication

*The Calf and Yearling
in 4-H Dairying*
PNW 78 — Pacific Northwest
Extension Publication

Feeding the Dairy Herd
North Central Regional
Extension Publication 346

*Feeding and Management of the
Dairy Calf: Birth to 6 Months*
Circular ANR-609 — AL
Cooperative Extension Service
Auburn, AL 36849-5612

*Feeding and Management
of Dairy Heifers: 6 Months
to Calving*
Circular ANR-632 — AL
Cooperative Extension Service

*Kansas Beef Leader
Curriculum Notebook*
State 4-H Office
KS State University
Manhattan, KS 66506-3404

Raising Dairy Replacements
North Central Regional
Extension Publication #205
University of WI, Madison

Skills for Life Beef Series
MN Extension Service
University of Minnesota
St. Paul, MN 55108-6069

Skills for Life Dairy Series
MN Extension Service

Working with Dairy Cattle
Holstein Foundation
P.O. Box 816
Brattleboro, VT 05302-0816

Wyoming 4-H Beef Manual
College of Agriculture
University of WY
P.O. Box 3354
University Station
Laramie, WY 82071-3354

Helpful Sources

STATE 4-H PROGRAMS

Contact your county Extension agent to find the 4-H program located near you.

DAIRY BREED ASSOCIATIONS

American Guernsey Assoc.
7614 Slate Ridge Blvd.
Reynoldsburg, OH 43068
(614-864-2409)
www.usguernsey.com

American Jersey Cattle Assoc.
6486 E. Main Street
Reynoldsburg, OH 43068-2362
(614-861-3636)
www.usjersey.com

American Milking Devon Assoc.
135 Old Bay Road
New Durham, NH 03855
(603-859-6611)
www.milkingdevons.org

American Milking Shorthorn Society
800 Pleasant St.
Beloit, WI 53511
(608-365-3332)
www.milkingshorthorn.com

Ayrshire Breeders' Assoc.
P.O. Box 1608
Brattleboro, VT 05302-1608
(802-254-7460)

Brown Swiss Cattle Breeders' Assoc.
800 Pleasant Street
Beloit, WI 53511-5456
(608-365-4474)
www.brownswissusa.com

Holstein Assoc. of America
One Holstein Place
Brattleboro, VT 05302-0808
(800-952-5200)
www.holsteinusa.com

Red & White Dairy Cattle Assoc.
3805 S. Valley Road
Crystal Spring, PA 15536
(814-735-4221)
www.redandwhitecattle.com

American Angus Association
3201 Frederick Ave.
St. Joseph, MO 64506
(816-383-5100)
www.angus.org

Beefmaster Breeders United
6800 Park Ten Blvd.
Suite 290 West
San Antonio, TX 78213
(210-732-3132)
www.beefmasters.org

**American Brahman
Breeders Assoc.**
3003 S. Loop West, Suite 140
Houston, TX 77054
(713-349-0854)
www.brahman.org

**International Brangus
Breeders Assoc.**
P.O. Box 696020
San Antonio, TX 78269-6020
(210-696-4343)
www.int-brangus.org

Braunvieh Assoc. of America
3815 Touzalin Ave., Suite 103
Lincoln, NE 68507
(402-466-3292)
www.braunvieh.org

**American-International
Charolais Assoc.**
P.O. Box 20247
Kansas City, MO 64195
(816-464-5977)
www.charolaisusa.com

American Chianina Assoc.
P.O. Box 890
Platte City, MO 64079
(816-431-2808)
www.chicattle.org

**American Dexter
Cattle Assoc.**
4150 Marino Ave.
Watertown, MN 55388
(952-446-1423)
www.dextercattle.org

**American Galloway
Breeders' Assoc.**
310 W. Spruce
Missoula, MT 59802
(406-728-5719)
www.americangalloway.com

American Gelbvieh Assoc.
10900 Dover Street
Westminster, CO 80021
(303-465-2333)
www.gelbvieh.org

American Hereford Assoc.
(includes Polled Herefords)
P.O. Box 014059
Kansas City, MO 64101
(816-842-3757)
www.hereford.org

North American
Limousin Foundation
7383 S. Alton Way, Suite 100
Englewood, CO 80112
(303-220-1693)
www.nalf.org

American Maine Anjou Assoc.
P.O. Box 1100
Platte City, MO 64079-1100
(816-431-9950)
www.maine-anjou.org

American Murray Grey Assoc.
P.O. Box 60748
Reno, NV 89506
(775-972-7526)
www.murraygreybeefcattle.com

American Pinzgauer Assoc.
P.O. Box 147
Bethany, MO 64424
(800-914-9883)
www.pinzgauers.org

Red Angus Assoc. of America
4201 I-35 North
Denton, TX 76207
(940-387-3502)
www.redangus.org

American Red Poll Assoc.
P.O. Box 147
Bethany, MO 64424
(660-425-7318)
www.redpollusa.org

American Romagnola Assoc.
3815 Touzalin, Suite 104
Lincoln, NE 68507
(402-466-3334)
www.americanromagnola.com

American Salers Assoc.
19590 E. Mainstreet #202
Parker, CO 80138
(303-770-9292)
www.salersusa.org

Santa Gertrudis Breeders
International
P.O. Box 1257
Kingsville, TX 78364
(361-592-9357)
www.santagertrudis.ws

American Shorthorn Assoc.
8288 Hascall Street
Omaha, NE 68124
(402-393-7200)
www.shorthorn.org

American Simmental Assoc.
Two Simmental Way
Bozeman, MT 59718
(406-587-2778)
www.simmgene.com

American Tarentaise Assoc.
P.O. Box 34705
N. Kansas City, MO 64116
(816-421-1993)
www.usa-tarentaise.com

Texas Longhorn Breeders
Assoc. of America
P.O. Box 4430
Fort Worth, TX 76164
(817-625-6241)
www.tlbaa.org

OTHER ORGANIZATIONS

National 4-H Council
7100 Connecticut Avenue
Chevy Chase, MD 20815
(301-961-2800)
www.fourhcouncil.edu

National Cattlemen's
Beef Assoc.
9110 E. Nichols Ave., #300
Centennial, CO 80112
(303-964-0305)
www.beef.org

University of Wisconsin
Center for Dairy Profitability
1675 Observatory Drive
266 Animal Science Bldg.
Madison, WI 53706
(608-263-5665)
http://cdp.wisc.edu

*"AgVentures" is a management
education program run through the
Center for Dairy Profitability that
offers instruction on how to manage a dairy farm.*

National Dairy Council
10255 West Higgins Rd
Suite 900
Rosemont, IL 60018
(847-803-2000)
www.nationaldairycouncil.org

National Dairy Herd
Improvement Assoc.
3021 E. Dublin Granville Rd
Suite 102
Columbus, OH 43231
(614-890-3630)
www.dhia.org

National FFA Center
P.O. Box 68960
Indianapolis, IN 46268-0960
(317-802-6060)
www.ffa.org

4-H Beef Project
Pacific Northwest Extension
Publication PNW 448
Oregon State University

4-H Dairy Cows and Management
North Central Regional
Extension Publication

*Buying, Caring for, Showing
Your Angus Heifer*
American Angus Association
St. Joseph, MO 64506

*The Calf and Yearling
in 4-H Dairying*
Pacific Northwest Extension
Publication PNW 78

*Early Heifer Development and
Colostrum Management*
Virginia Cooperative Ext.
Dairy Science Pub. 404-282

Feeding the Dairy Herd
North Central Regional
Extension Publication #346
University of Minnesota

*Feeding and Management of the
Dairy Calf: Birth to 6 Months*
Alabama Cooperative
Ext. Service Circular ANR-609
Auburn, AL 36849-5612

*Feeding and Management of Dairy
Heifers: 6 Months to Calving*
Alabama Cooperative Ext.
Service Circular ANR-632

*Feeding Dairy Cows for Efficient
Reproductive Performance*
North Central Regional Ext.
Publication #366

Flat-Barn Milking Systems
University of Wisconsin Ext.
Publication A3567

*Improving the Detection of
Estrus in Dairy Cattle*
Purdue University Cooperative
Extension Service #AS-453
West Lafayette, IN

*Kansas Beef Leader
Curriculum Notebook*
State 4-H Office
Kansas State University
Manhattan, KS 66506-3404

*Manual Cleaning of
Milking Equipment*
University of Wisconsin
Extension Publication A1300

*Nutrition for the Early
Developing Heifer*
Virginia Cooperative Extension
Dairy Science Pub. 404-283

Raising Dairy Replacements
North Central Regional Ext.
Publication #205
U. of Wisconsin, Madison

Skills for Life Beef Series
Minnesota Extension Service
University of Minnesota
St. Paul, MN 55108-6069

Skills for Life Dairy Series
Minnesota Extension Service
University of Minnesota

*Trouble-Shooting High Bacteria
Counts in Farm Milk*
University of Wisconsin Ext.
Publication A3705

*Updated Nutrient Specifications
for the Dairy Herd*
Virginia Cooperative Extension
Dairy Publication 404-105

Using AI Young Sires in Your Herd
University of Wisconsin Ext.
Publication A3559

Working With Dairy Cattle
Holstein Foundation
P.O. Box 816
Brattleboro, VT 05302-0816
(800-952-5200)

Wyoming 4-H Beef Manual
College of Agriculture
University of Wyoming
P.O. Box 3354
University Station Laramie, WY
82071-3354

*Youth and Dairy Cattle:
A Safe Partnership*
University of Wisconsin Ext.
Publication A3705

Identify Your Cattle
University of Wisconsin Ext.
Publication A2834

A list of other helpful dairy publications can be obtained from the
University of Wisconsin's Cooperative Extension Publications, Room B-8,
45 N. Charter St., Madison, WI 53715 (608-262-3346 or 877-947-7827)
www.uwex.edu/ces/pubs

Most of the breed associations publish a magazine or newsletter, and these can be very helpful. If you want to raise purebreds, or even if you are just interested in a certain breed, contact that association to find about their publications. You can also find some regional publications geared specifically to agriculture/ livestock production in your local area.

American Red Angus
4201 North I-35
Danton, TX 76207-3415
(940-387-3502)
www.redangus.org

American Small Farm
267 Broad Street
Westerville, OH 43081
(614-895-3755)
www.smallfarm.com

Angus Beef Bulletin
3201 Frederick Ave.
St. Joseph, MO 64506
(800-821-5478)
www.angusbeefbulletin.com

BEEF
7900 International Drive
Suite 300
Minneapolis, MN 55425
(952-851-4710)
www.beef-mag.com

Beef Today
1818 Market Street, 31st Floor
Philadelphia, PA 19103
(800-523-1538)
www.agweb.com

Cascade Cattleman
P.O. Box 1390
Klamath Falls, OR 97601
(800-275-0788)
www.cascadecattleman.com

Cattle Today
204 S. Temple Ave.
Fayette, AL 35555
(205-932-8000)
www.cattletoday.com

The Cattleman
1301 W. Seventh St.
Fort Worth, TX 76102
(817-332-7155)

Cow Country News
176 Pasadena Drive
Lexington, KY 40503
(859-278-0899)
kycattle.org/cowcountry.cfm

Dairy Herd Management
10901 West 84th Terrace
Lenexa, KS 66214
(800-255-5113)
www.dairyherd.com

Dairy Today
1818 Market Street, 31st Floor
Philadelphia, PA 19103
(800-523-1583)
www.agweb.com

Drovers
10901 West 84th Terrace
Lenexa, KS 66214
(800-255-5113)
www.drovers.com

The Fence Post
421 Main Street
Windsor, CO 80550
www.thefencepost.com

GRAINEWS
P.O. Box 9800
Winnipeg, MB R3C 3K7
Canada
www.agcanada.com

Hereford World
P.O. Box 014059
Kansas City, MO 64101
(816-842-3757)
www.hereford.org

Hobby Farms
P.O. Box 58701
Boulder, CO 80322-8701
(800-365-4421)
www.hobbyfarmsmagazine.com

Limousin World
2005 Ruhl Drive
Guthrie, OK 73044
(405-260-3775)
www.limousinworld.com

Progressive Farmer
P.O. Box 2581
Birmingham, AL 35202
(800-357-4466)
www.progressivefarmer.com

Rural Heritage
281 Dean Ridge Lane
Gainesboro, TN 38562-5039
(931-268-0655)
www.ruralheritage.com

Small Farmers Journal
P.O. Box 1627
Sisters, OR 97759
(541-549-2064)
www.smallfarmersjournal.com

Small Farm Today
3903 W. Ridge Trail Road
Clark, MO 65243
(573-687-3525)
www.smallfarmtoday.com

Southern Livestock Review
P.O. Box 423
Somerville, TN 38068
(901-465-4042)
www.southernlivestockrev.com

The Stockman Grass Farmer
P.O. Box 2300
Ridgeland, MS 39158
(800-748-9808)
www.stockmangrassfarmer.com

Texas Longhorn Trails
P.O. Box 4430
Fortworth, TX 76164
(817-625-6241)
www.tlbaa.org

Tri-State Livestock News
P.O. Box 129
Sturgis, SD 57785
(800-253-3656)
www.tsln.com

Western Cowman
P.O. Box 613
Fair Oaks, CA 95628
(916-362-2697)

INFORMATION ON RAISING NATURAL, GRASS-FED & ORGANIC BEEF

www.alderspring.com
A rancher who produces natural beef

http://dare.agsci.colostate.edu/aftnichebeef
A Web site put together by Colorado State University and American
Farmland Trust, covering various aspects of niche marketing

www.countrynaturalbeef.com
An example of a cooperative that markets natural beef

www.westonaprice.org
A good source for information on raw milk

www.albc-usa.org
Web site for the American Livestock Breeds Conservancy, an organization dedicated to protecting rare breeds

Glossary

abomasum *(n.)* A compartment of the stomach.

abortion *(n.)* Loss of pregnancy; dead fetus expelled early.

abscess *(n.)* Pus-filled swelling.

acidosis *(n.)* Severe digestive upset caused by too much grain.

afterbirth *(n.)* (Placenta) tissue encasing the calf, attached to the uterus; it comes out after the calf.

amnion *(n.)* Membrane enclosing the calf when he is born.

antibiotic *(n.)* Drug used to combat bacterial infections.

antibodies *(n.)* Protein molecules in the bloodstream that fight a specific disease.

artificial insemination (AI) *(n.)* Process in which a technician puts semen from a bull into the cow's uterus to create pregnancy.

backgrounding *(v.)* Feeding weaned calves for awhile on pasture or crop aftermath until they are older or larger and ready to go into a feedlot for finishing on grain.

bacteria *(n.)* Tiny one-celled organisms; some cause disease.

balance *(n.)* Harmonious relationship of all body parts.

balanced ration *(n.)* Daily food in the right mixtures and amounts to include all required nutrients.

balling gun *(n.)* Tool for placing a pill or bolus at the back of the mouth to make a cow swallow it.

bang's disease *(n.)* Brucellosis; causes abortion in cows and undulant fever in humans.

birth canal *(n.)* Vagina; where the calf comes out from the uterus.

birth weight *(n.)* What the calf weighs when he is born.

blackleg *(n.)* Serious disease of cattle caused by soil bacteria.

bloat *(n.)* Tight, swollen rumen (left side when viewed from behind) caused by accumulation of gas.

bos indicus *(n.)* Species of humped cattle common to the tropics.

bos taurus *(n.)* Species of cattle originating in cooler regions.

bovine *(n., adj.)* Term referring to cattle.

bred *(adj.)* Mated, pregnant.

breed *(n.)* Group of animals that have the same ancestry and characteristics.

breeding *(n.)* When a heifer or cow is mated with a bull. Family history.

brucellosis *(n.)* Bacterial disease (Bang's) that causes abortion.

bucket calf *(n.)* Calf fed milk from a bucket.

bull *(n.)* Uncastrated male of any age.

bulk tank *(n.)* Large stainless-steel refrigerator tank for cooling and storing fresh milk.

bunt *(v.)* Hitting with the head (as a calf bunting the cow's udder).

butt *(v.)* Hitting with the head (as a cow or calf hitting you or another animal with its head).

butterfat *(n.)* Fat content in milk that can be separated out to become cream or butter.

BVD *(n.)* Bovine Viral Diarrhea, a serious viral infection of cattle that can cause illness and abortion.

by-product *(n.)* Something made from leftover parts; leather is a by-product of cattle that are butchered for meat.

calf *(n.)* Young animal, either sex, less than a year old.

calving *(v.)* Giving birth to a calf.

carotene *(n.)* Orange-colored plant material that serves as source of vitamin A.

castrate *(v.)* To remove the testicles of male cattle.

cervix *(n.)* Opening (or seal) between uterus and vagina.

cesarean section *(n.)* Delivery of a calf through surgery.

CLA *(n.)* Conjugated Linoleic Acid, a fatty acid found in beef and dairy foods that helps promote healthy cell structure.

clostridial diseases *(n.)* Deadly diseases caused by spore-forming bacteria that exist a long time in the environment. These include tetanus, blackleg, malignant edema, redwater, enterotoxemia.

coccidiosis *(n.)* Intestinal disease and diarrhea caused by protozoans.

colostrum *(n.)* First milk after a cow calves; contains antibodies that give the calf temporary protection against certain diseases.

commercial cattle *(n.)* Unregistered, not purebred, raised for beef.

composite *(n.)* Uniform group of cattle created by selective crosses.

conceive *(v.)* Become pregnant.

concentrates *(n.)* Feeds low in fiber and high in food value; grains.

conformation *(n.)* General structure and shape of an animal.

contagious *(adj.)* Readily transmitted from one animal to another.

continental breed *(n.)* Originating in Europe rather than the British Isles.

cow *(n.)* Bovine female that has had one or more calves.

cow hocked *(adj.)* Hind legs too close together at the hocks.

crossbred *(n.)* Animal resulting from crossing two or more breeds.

cud *(n.)* Wad of food regurgitated from the rumen to be rechewed.

cull *(v.)* Eliminate (sell) an animal of low quality from the herd.

cycling *(v.)* Having heat cycles.

dam *(n.)* Mother of the calf.

dehorn *(v.)* Remove horns from an animal.

dewclaw *(n.)* Horny structure on the lower leg, above the hoof.

dewlap *(n.)* Loose skin under the neck.

deworm *(v.)* Treat with a product to kill internal parasites.

digestion *(n.)* Process of breaking down feeds into nutrients.

diphtheria *(n.)* Bacterial disease of calves in the mouth and throat.

disposition *(n.)* Temperament and attitude.

dry period *(n.)* The time a cow is not producing milk.

dual purpose *(adj.)* Usable as a breed for both milk and meat.

electrolytes *(n.)* Important body salts.

EPD *(n.)* Expected Progeny Difference, an estimate of how much better or poorer an animal's offspring will perform compared to the average of all the individuals in the herd or breed.

estrus *(n.)* Heat period; when the cow will accept the bull.

exotic *(n.)* Recently imported breed (European breed).

feed conversion *(n.)* Ratio of pounds of feed eaten to pound of gain.

feedlot *(n.)* Pen where cattle are fattened.

femininity *(n.)* Female characteristics such as udder development, refinement of body, head, neck, and so on.

fertility *(n.)* Ability to reproduce.

fetus *(n.)* Developing calf (in the uterus) after 45 days of pregnancy.

fever *(n.)* Body temperature above normal.

fiber *(n.)* Coarse part of the feed, not easily digested.

finish *(v.)* Become fat enough to be butchered.

fitting *(v.)* Clipping, washing, and brushing an animal for show.

flight zone *(n.)* Distance you can get to an animal before it flees.

footrot *(n.)* Infection (from soil bacteria) causing severe lameness.

forage *(n.)* Pasture and hay; roughages.

founder *(n.)* Inflammation of the hoofs caused by overfeeding grain.

frame score *(n.)* Measure of hip or shoulder height to determine skeletal size.

freshen *(v.)* Give birth to a calf and begin producing milk.

fungi *(n.)* Primitive parastic plants that reproduce by spores.

genetics *(n.)* Qualities and physical characteristics that are inherited.

gestation *(n.)* Length of pregnancy (about 285 days for cattle).

grade *(n.)* Unregistered, not purebred.

grass-fed beef *(n.)* Meat from animals raised predominantly on forages, with little or no concentrates (grains or supplements).

grass-finished beef *(n.)* Meat from animals raised entirely on forages, with no concentrates (grain or supplements) in the diet.

growth implants *(n.)* Artificial hormones given to young beef animals to stimulate more rapid growth.

grub *(n.)* Immature stage (larvae) of the heel fly that overwinters in the bodies of cattle as an internal parasite.

gut *(n.)* Digestive tract.

halter *(n.)* Rope or strap looped behind the ears and around the nose to control or lead the animal.

hardware disease *(n.)* Peritonitis (infection in the abdomen) caused by a sharp foreign object penetrating the reticulum wall.

headcatcher *(n.)* The gated end of a cattle chute, used to restrain cattle by immobilizing the head.

heat *(n.)* (See estrus.)

heifer *(n.)* Young female, before she has a calf.

herbicide *(n.)* Product that kills certain plants, such as weeds.

heredity *(n.)* Transmission of characteristics from parents to offspring.

hobble *(v.)* Tie the legs together.

hybrid vigor *(n.)* The degree to which a crossbred offspring outperforms his purebred parents.

IBR *(n.)* Infectious Bovine Rhinotracheitis, a serious viral infection of cattle that causes respiratory problems and abortion.

immunity *(n.)* Ability to resist a certain disease.

infection *(n.)* Invasion of body tissues by germs or parasites.

insecticide *(n.)* Product that kills insects.

interest *(n.)* Fee paid on money borrowed.

intramuscular (IM) *(adj.)* Into the muscle (as an injection).

intravenous (IV) *(adj.)* Into a vein.

iodine *(n.)* Harsh chemical used for disinfecting.

ionophere *(n.)* An agent that combines with an ion and transports it across a membrane. Certain antibiotics fed to cattle at low levels in grain, mineral, or bolus are ionophores, enhancing feed efficiency and helping to prevent bloat.

labor *(n.)* The cow's efforts in pushing the calf out at birth.

lactation *(n.)* Producing milk.

legume *(n.)* Plant belonging to the pea family (alfalfa, clover).

leptospirosis *(n.)* Bacterial disease that can cause abortion.

lice *(n.)* Tiny external parasites on the skin.

lump jaw *(n.)* Abscess caused by infection in the mouth.

malignant edema *(n.)* A fatal infection of cattle caused by clostridial bacteria.

marbling *(n.)* Flecks of fat interspersed in muscle (beef).

market value *(n.)* Price received for an animal.

mastitis *(n.)* Infection and inflammation in the udder.

maternal traits *(n.)* Characteristics that make a good cow.

monensin *(n.)* The generic name for a type of veterinary antibiotic administered to cattle at low levels to protect them from certain diseases and to keep them from bloating.

muzzle *(n.)* Nose and mouth.

natural beef *(n.)* Meat from animals raised without the use of antibiotics or growth-stimulating hormones.

navel *(n.)* Area where the umbilical cord was attached.

niche market *(n.)* Fraction of a larger market, with special criteria for the product that make it desired by certain consumers.

omasum *(n.)* One of the four stomach compartments.

omega-3 fatty acids *(n.)* The "good" (healthful) fatty acids in foods.

omega-6 fatty acids *(n.)* The "bad" fatty acids in food.

open *(adj.)* Not pregnant.

organic beef *(n.)* Meat from cattle raised under natural conditions, using management methods that include no contact with human-made chemicals except for vaccines.

ovary *(n.)* Female reproductive gland where eggs are formed.

parasite *(n.)* Organism that lives in or on an animal.

pasteurize *(v.)* To heat milk to a certain temperature to kill germs. Named after Louis Pasteur, who discovered germs and how to kill them with heat.

pedigree *(n.)* Chart of the ancestors of an animal.

penis *(n.)* The male organ that passes sperm into the female and also passes urine.

PI3 *(n.)* Parainfluenza 3, a viral disease of cattle that can cause respiratory problems.

pin bones *(n.)* Bony part of pelvis that protrudes on either side of the rectum.

pinkeye *(n.)* Contagious eye infection spread by face flies.

placenta *(n.)* Afterbirth; attached to the uterus during pregnancy.

pneumonia *(n.)* Infection in the lungs.

polled *(adj.)* Born without horns; naturally hornless.

production records *(n.)* Measure of milk produced, or calf weaning weights, and so on.

progeny *(n.)* Offspring.

protein supplement *(n.)* Concentrate containing 32–44 percent protein.

puberty *(n.)* Age when the animal matures sexually and can reproduce.

purebred *(n.)* Member of a certain breed (such as Hereford and Angus), with no other breeds in its ancestry. Not to be confused with Thoroughbred (a breed of horse).

quarter *(n.)* One of four compartments of the cow's udder.

raw milk *(n.)* Milk straight from the cow, not pasteurized.

redwater *(n.)* Deadly bacterial disease of cattle.

registered *(v.)* Recorded in the herd book of a breed.

retained placenta *(n.)* The cow fails to shed the placenta quickly.

reticulum *(n.)* One of the cow's four stomachs.

ringworm *(n.)* Fungal infection causing scaly patches of skin.

roughages *(n.)* Feeds high in fiber and low in energy (hay, pasture).

rumen *(n.)* Largest stomach compartment, where roughage is digested.

rumensin *(n.)* (Monensin) A type of veterinary antibiotic often used in feed at low levels to protect cattle from certain diseases and to keep them from bloating.

ruminant *(n.)* Animal that chews its cud and has a four-part stomach.

scours *(n.)* Diarrhea; can be caused by infection or improper feed.

scrotum *(n.)* Sac enclosing the testicles of a bull.

semen *(n.)* Fluid put forth by the bull (containing sperm) when breeding a cow.

settle *(v.)* Become pregnant.

sheath *(n.)* Tube-shaped fold of skin into which the penis retracts.

sire *(n.)* Father of a calf.

somatic cells *(n.)* Cells in the milk, indicating udder infection.

splay footed *(adj.)* Feet toe out.

stanchion *(n.)* Upright frame for confining a cow by the head; she puts her head through and it closes on her neck so she can't back out until it is opened again.

steer *(n.)* Male bovine, after castration.

stifle *(n.)* Large joint, high on the hind leg, by the flank.

straightbred *(n.)* Animal with parents of the same breed, but not necessarily purebred.

straw *(n.)* Stems of plants grown for grain; often used as bedding.

stress *(n.)* Abnormal or adverse conditions that are hard on an animal.

subcutaneous (SQ) *(adj.)* Directly under the skin (as an injection).

supplement *(n.)* Addition (vitamins, minerals, or protein) to the diet that makes the total ration more complete or balanced.

switch *(n.)* End of the tail, where the hair is longest.

tattoo *(n.)* Permanent mark in the ear.

teat *(n.)* The "nipple" on each quarter of the udder.

testes, testicles *(n.)* Male reproductive glands.

udder *(n.)* Mammary glands and teats.

umbilical cord *(n.)* Attaches the calf to the placenta and uterus.

uterine contractions *(n.)* Waves in the muscle (like a swallowing motion) that push the calf out of the uterus.

uterus *(n.)* Portion of the reproductive tract where the calf develops.

vaccinate *(n.)* Administering a vaccine.

vaccine *(n.)* Fluid containing killed or modified germs, put into an animal's body to stimulate production of antibodies and immunity.

vagina *(n.)* Tube into the uterus from the vulva.

virus *(n.)* Tiny particle that invades cells to cause disease.

vulva *(n.)* External opening of the vagina.

warble *(n.)* Larva of the heel fly; it burrows out through the skin on the cow's back.

warts *(n.)* Skin growths caused by a virus.

waste milk *(n.)* Milk that can be used for calves but not for human food, including colostrum, transitional milk (milk that still contains some colostrum), and milk from a cow with mastitis or that may contain antibiotics after treating a cow for mastitis.

water sac *(n.)* Fluid-filled membrane that breaks during birth.

wean *(v.)* To separate a calf from his mother or stop feeding him milk.

weanling *(n.)* Recently weaned calf (up to a year of age).

yearling *(n.)* Calf between one and two years of age.

Index

Page references in *italics* indicate illustrations; **bold** indicates charts.

Other Storey Titles You Will Enjoy

Basic Butchering of Livestock and Game by John J. Mettler, Jr., DVM.
Whether you hunt or work a small farm and hope to become more self-sufficient, this book can help you enjoy the better flavor of humanely and properly slaughtered and butchered meat. Detailed instructions and easy-to-follow illustrations demystify the slaughtering and butchering process. 208 pages. Paperback. ISBN 0-88266-391-7.

The Family Cow by Dick van Loon.
This is the basic book — yet with all the essential details — for the family that decides to keep a cow for all the benefits she can provide. Practical, fully-illustrated chapters cover every topic, from the history of the cow to handling techniques. 272 pages. Paperback. ISBN 0-88266-066-7.

How to Build Animal Housing by Carol Ekarius.
This book helps you evaluate the housing needs of your animals and provides dozens of adaptable plans for sheds, coops, hutches, multi-purpose barns, windbreaks, and shade structures, as well as plans for essential equipment. You'll also get tried-and-true advice on the importance of planning ahead and budgeting adequately. 272 pages. Paperback.
ISBN 1-58017-527-9.

Small-Scale Livestock Farming by Carol Ekarius.
Small farms can pay big dividends, Ekarius explains, but success demands knowledge and effective management. Through case studies of successful farmers, nitty-gritty details on every facet of livestock farming, and fascinating insights for working with nature instead of against it, you'll learn to make your farm thrive. 224 pages. Paperback. ISBN 1-58017-162-1.

Storey's Guide to Raising Beef Cattle by Heather Smith Thomas.
Whether you want to raise one or two animals or run a full-scale beef production operation, this definitive handbook has all the information you need. 352 pages. Paperback. ISBN 1-58017-327-6.

Storey's Guide to Raising Dairy Goats by Jerry Belanger.
This indispensable, fully-illustrated guide provides the very latest practical information for dairy goat owners. 288 pages. ISBN 1-58017-259-8.